Models of Horizontal Eye Movements

Part II: A 3rd Order Linear Saccade Model

Models of Horizontal Eye Movements, Part II: A 3rd Order Linear Saccade Model

John D. Enderle and Wei Zhou

ISBN: 978-3-031-00515-2   paperback
ISBN: 978-3-031-01643-1   ebook

DOI 10.1007/978-3-031-01643-1

A Publication in the Springer series
SYNTHESIS LECTURES ON BIOMEDICAL ENGINEERING

Lecture #35
Series Editor: John D. Enderle, *University of Connecticut*
Series ISSN
Synthesis Lectures on Biomedical Engineering
Print 1930-0328   Electronic 1930-0336

Synthesis Lectures on Biomedical Engineering

Editor

John D. Enderle, *University of Connecticut*

Lectures in Biomedical Engineering will be comprised of 75- to 150-page publications on advanced and state-of-the-art topics that spans the field of biomedical engineering, from the atom and molecule to large diagnostic equipment. Each lecture covers, for that topic, the fundamental principles in a unified manner, develops underlying concepts needed for sequential material, and progresses to more advanced topics. Computer software and multimedia, when appropriate and available, is included for simulation, computation, visualization and design. The authors selected to write the lectures are leading experts on the subject who have extensive background in theory, application and design.

The series is designed to meet the demands of the 21st century technology and the rapid advancements in the all-encompassing field of biomedical engineering that includes biochemical, biomaterials, biomechanics, bioinstrumentation, physiological modeling, biosignal processing, bioinformatics, biocomplexity, medical and molecular imaging, rehabilitation engineering, biomimetic nano-electrokinetics, biosensors, biotechnology, clinical engineering, biomedical devices, drug discovery and delivery systems, tissue engineering, proteomics, functional genomics, molecular and cellular engineering.

Models of Horizontal Eye Movements, Part II: A 3rd Order Linear Saccade Model
John D. Enderle and Wei Zhou
2010

Models of Horizontal Eye Movements, Part I: Early Models of Saccades and Smooth Pursuit
John D. Enderle
2010

Biomedical Technology Assessment: The 3Q Method
Phillip Weinfurt
2010

Strategic Health Technology Incorporation
Binseng Wang
2009

Advanced Probability Theory for Biomedical Engineers
John D. Enderle, David C. Farden, Daniel J. Krause
2006

Intermediate Probability Theory for Biomedical Engineers
John D. Enderle, David C. Farden, Daniel J. Krause
2006

Basic Probability Theory for Biomedical Engineers
John D. Enderle, David C. Farden, Daniel J. Krause
2006

Sensory Organ Replacement and Repair
Gerald E. Miller
2006

Artificial Organs
Gerald E. Miller
2006

Signal Processing of Random Physiological Signals
Charles S. Lessard
2006

Image and Signal Processing for Networked E-Health Applications
Ilias G. Maglogiannis, Kostas Karpouzis, Manolis Wallace
2006

Models of Horizontal Eye Movements

Part II: A 3rd Order Linear Saccade Model

John D. Enderle
University of Connecticut

Wei Zhou
University of Connecticut

SYNTHESIS LECTURES ON BIOMEDICAL ENGINEERING #35

ABSTRACT

There are five different types of eye movements: saccades, smooth pursuit, vestibular ocular eye movements, optokinetic eye movements, and vergence eye movements. The purpose of this book is focused primarily on mathematical models of the horizontal saccadic eye movement system and the smooth pursuit system, rather than on how visual information is processed. A saccade is a fast eye movement used to acquire a target by placing the image of the target on the fovea. Smooth pursuit is a slow eye movement used to track a target as it moves by keeping the target on the fovea. The vestibular ocular movement is used to keep the eyes on a target during brief head movements. The optokinetic eye movement is a combination of saccadic and slow eye movements that keeps a full-field image stable on the retina during sustained head rotation. Each of these movements is a conjugate eye movement, that is, movements of both eyes together driven by a common neural source. A vergence movement is a non-conjugate eye movement allowing the eyes to track targets as they come closer or farther away.

In this book, a 2009 version of a state-of-the-art model is presented for horizontal saccades that is 3rd-order and linear, and controlled by a physiologically based time-optimal neural network. The oculomotor plant and saccade generator are the basic elements of the saccadic system. The control of saccades is initiated by the superior colliculus and terminated by the cerebellar fastigial nucleus, and involves a complex neural circuit in the mid brain. This book is the second part of a book series on models of horizontal eye movements.

KEYWORDS

saccade, main sequence, time-optimal control, system identification, dynamic overshoot, glissadic overshoot, neural network

Contents

Preface

There are five different types of eye movements: saccades, smooth pursuit, vestibular ocular eye movements, optokinetic eye movements, and vergence eye movements. The purpose of this book is focused primarily on mathematical models of the horizontal saccadic eye movement system and the smooth pursuit system, rather than on how visual information is processed. A saccade is a fast eye movement used to acquire a target by placing the image of the target on the fovea. Smooth pursuit is a slow eye movement used to track a target as it moves by keeping the target on the fovea. The vestibular ocular movement is used to keep the eyes on a target during brief head movements. The optokinetic eye movement is a combination of saccadic and slow eye movements that keeps a full-field image stable on the retina during sustained head rotation. Each of these movements is a conjugate eye movement, that is, movements of both eyes together driven by a common neural source. A vergence movement is a non-conjugate eye movement allowing the eyes to track targets as they come closer or farther away.

In this book, a 2009 version of a state-of-the-art model is presented for horizontal saccades that is 3rd-order and linear, and controlled by a physiologically based time-optimal neural network. The oculomotor plant and saccade generator are the basic elements of the saccadic system. The control of saccades is initiated by the superior colliculus and terminated by the cerebellar fastigial nucleus, and involves a complex neural circuit in the mid brain. This book is the second part of a book series on models of horizontal eye movements.

The oculomotor plant consists of the eyeball, the lateral and medial rectus muscle, and a Voigt element (a pair of viscosity and elasticity elements in parallel) representing the passive restraining effect of orbital tissues and the other four muscles. The rectus muscle is modeled with a Voigt element in series and another Voigt element in parallel with an active state tension generator. This linear muscle model exhibits accurate nonlinear force-velocity and length-tension relationships. The oculomotor plant is 3rd-order.

The signals that drive saccades are described by pulse-slide-step waveforms, with a post inhibitory rebound burst and based on a physiologically based time-optimal controller. An anatomically based neural network is presented whereby the saccade is initiated by the Superior Colliculus and terminated by the Cerebellum. As a natural progression from the model, a new theory describing saccades with dynamic overshoot and glissadic overshoot is presented. System identification is presented that allows for estimating the neural controller and the parameters of the oculomotor plant.

The work presented here is not an exhaustive coverage of the field, but focused on the interests of the authors. We wish to express my thanks to William Pruehsner for drawing many of the illustrations in this book, and Kerrie Wenzler and Gresa Ajetifor editorial assistance.

John D. Enderle and Wei Zhou
March 2010

CHAPTER 1

2009 Linear Homeomorphic Saccadic Eye Movement Model and Post-Saccade Behavior: Dynamic and Glissadic Overshoot[1]

1.1 INTRODUCTION

Up to now, we have considered saccades that end their trajectory smoothly and without abrupt changes. Post-saccade phenomena, such as a dynamic or a glissadic overshoot, are usually observed during human saccades (Weber and Daroff, 1972). In dynamic overshoot, the eyes move beyond the target, and then, with a quick saccade-like return with no time delay, the eyes move back to the target. Glissadic overshoot is similar to dynamic overshoot but with a return that is slower.

Bahill and coworkers report that dynamic overshoot occurs in 70% of saccades in humans, and they are quite random; a subject on one day will have many saccades with dynamic overshoot, and on another very few (Bahill et al., 1975). Further, they report that undershoot is not common. Kapoula et al. (1986) report that dynamic overshoot occurs with a frequency of 13% during saccades, more frequent for saccades of 10° or less, and almost always in the abducting eye. Westine and Enderle (1988) report large variability in post-saccade behavior with five human subjects having dynamic overshoot in 5%-40% of their saccades and a glissadic overshoot in 22%-52% of their saccades.

Pulse-step mismatch provides one explanation for post-saccade phenomena (Quaia and Optican, 1998, 2003). Bahill et al. (1975) also describe a neural control reversal as an explanation for post-saccade phenomena. However, Enderle and coworkers suggest another explanation, that is, a post-inhibitory rebound burst (PIRB) in the antagonist motoneurons causes the post-saccade phenomena (Enderle and Wolfe, 1988; Enderle and Engelken, 1995; Enderle, J.,

---

[1]Some of the material in this chapter is an expansion of a previously published paper: Zhou et al. (2009). An Updated Time-Optimal 3rd-Order Linear Saccadic Eye Plant Model. *International Journal of Neural Systems,* Vol. 19, No. 5, 309-330 and the MS Thesis by Wei Zhou, A Mathematical Model For Horizontal Saccadic Eye Movement Based on a Truer Linear Muscle Model and Time-Optimal Controller, 2006.

2002). PIRB firing occurs when a neuron is profoundly inhibited, and when released from inhibition without stimulus, fire with a high-frequency burst that ends after a short period of time. PIRB is observed in a monkey agonist motoneuron when it first starts firing during a saccade (for example, see Fig. 2(A) at the arrow in Sylvestre and Cullen, 1999). PIRB is also observed in a monkey antagonist motoneuron when it resumes firing after the agonist pulse (for example, see Fig. 4 in Robinson, D., 1981 and Fig. 2 in Gisbergen et al., 1981). This behavior is also observed in humans (Kapoula et al., 1986). The PIRB observed in the motoneuron is actually caused by the excitatory burst neuron (EBN[2]) as described later.

In 1988, Enderle and Wolfe (Enderle and Wolfe, 1988), described using the system identification technique to estimate the parameters of a 4th order model of the oculomotor plant presented in Section 3.6 in Book 1 and the active state tensions during the saccadic eye movement. The active state tension is modeled as a low-pass filtered pulse-step waveform as described earlier. Parameter estimates are calculated for the model using a conjugate gradient search program that minimizes the integral of the absolute value of the squared error between the model and the data in the frequency domain. Initial parameter estimates are based on physiological evidence. For saccades that end without post-saccade behavior, the estimation results are in excellent agreement with the data as shown in Figs. 3.31–3.33 in Book 1.

In this chapter, we wish to further explore the fast eye movement system that has post-saccade behavior, including normal saccades and those with a dynamic or a glissadic overshoot based on a model by Zhou et al. (2009). To analyze post-saccade behavior, the neural input to the muscles is now described by a pulse-slide-step of neural activity, supported by physiological evidence (Goldstein, H., 1983). The slide is a slow exponential transition from the pulse to the step. We will also explore system identification to estimate parameters for this model and a time-optimal controller in this chapter.

1.2   OCULOMOTOR PLANT

The oculomotor plant is shown in Fig. 1.1. The eye muscles are identical to those in Section 5.2 of Book 1, where all elements are linear. This linear muscle model exhibits accurate nonlinear force-velocity and length-tension relationships. No other linear muscle model to date is capable of accurately reproducing these nonlinear relationships.

The inputs to the muscle model are the agonist and antagonist active-state tensions, which are derived from a low-pass filtering of the saccadic neural innervation signals. As previously mentioned, the neural innervation signals are typically characterized as a pulse-step signal or a pulse-slide-step signal during saccadic eye movement (Goldstein, H., 1983; Optican and Miles, 1985).

It should be noted that the passive elasticity and viscosity of the eyeball in Fig. 1.1 is changed from the model in Section 5.3 of Book 1 that included two Voigt passive elements connected in series to a single Voigt element. The Voigt element with time constant 0.02 s is used in the model presented here. The other Voigt element, with a time constant of 1 s, is neglected since it has an

---

[2]A table of abbreviations of the neural sites that generate a saccade is provided in Table 2.1.

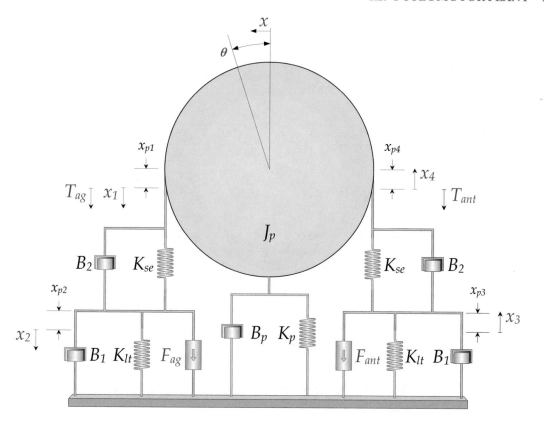

Figure 1.1: Oculomotor plant used for analyzing saccades with post-saccade behavior.

insignificant effect on the accuracy as we are modeling a single saccade and not a series of saccades. Further, eliminating this Voigt element reduces the order of the model from 4th to 3rd order and simplifies the system identification.

The net torque generated by the muscles during a saccade rotates the eyeball to a new orientation, and after the saccade is completed, compensates the passive restraining torques generated by orbital tissues.

1.2.1 DERIVATION OF THE DIFFERENTIAL EQUATION DESCRIBING THE OCULOMOTOR SYSTEM

To begin our analysis of this model, we assume:

1. $\dot{x}_2 > \dot{x}_1 = \dot{x} = \dot{x}_4 > \dot{x}_3$.

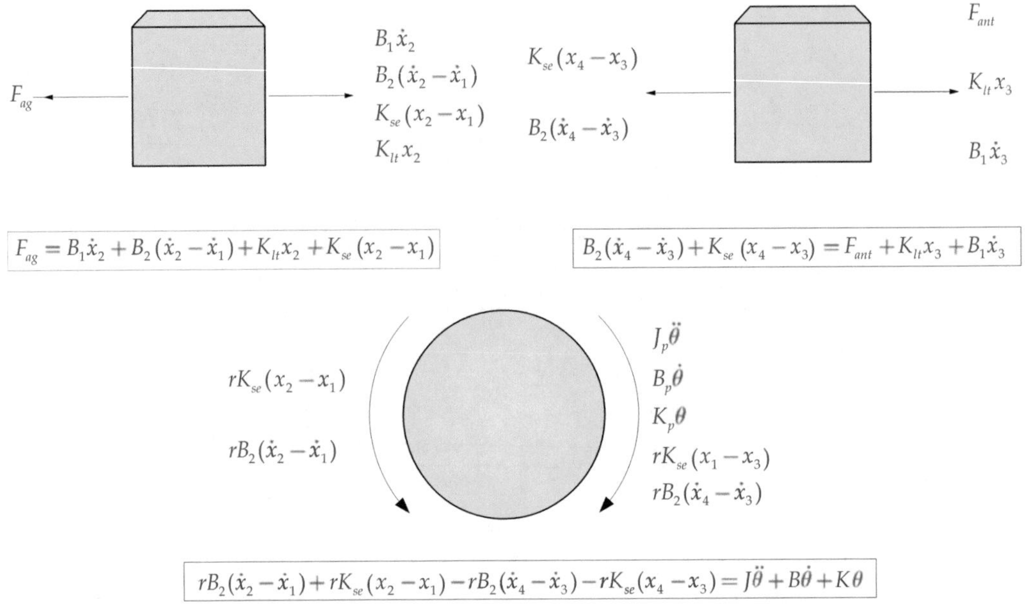

$$F_{ag} = B_1\dot{x}_2 + B_2\left(\dot{x}_2 - \dot{x}_1\right) + K_{lt}x_2 + K_{se}\left(x_2 - x_1\right)$$

$$B_2\left(\dot{x}_4 - \dot{x}_3\right) + K_{se}\left(x_4 - x_3\right) = F_{ant} + K_{lt}x_3 + B_1\dot{x}_3$$

$$rB_2\left(\dot{x}_2 - \dot{x}_1\right) + rK_{se}\left(x_2 - x_1\right) - rB_2\left(\dot{x}_4 - \dot{x}_3\right) - rK_{se}\left(x_4 - x_3\right) = J\ddot{\theta} + B\dot{\theta} + K\theta$$

Figure 1.2: Free body diagrams for system in Fig. 1.1.

2. Elasticity K_p is the passive elasticity for muscles other than the lateral and medical rectus muscle and the eyeball.

3. Viscous element B_p is due to the friction of the eyeball within the eye socket and for muscles other than the lateral and medical rectus muscle.

4. x_i is measured from the equilibrium position.

5. Zero initial conditions.

Note that $x = \theta r$ or $\theta = \frac{x}{r} \times \frac{180}{\pi} = 5.2087 \times 10^3 x$, where x is measured in meters with $r = 11$ mm.

The free body diagrams are shown in Fig. 1.2, which give the node equations for the system:

Node x:

$$r B_2(\dot{x}_2 - \dot{x}_1) + r K_{se}(x_2 - x_1) - r B_2(\dot{x}_4 - \dot{x}_3) - r K_{se}(x_4 - x_3) = J_p\ddot{x} + B_p\dot{x} + K_p x . \quad (1.1)$$

Node 2:

$$F_{ag} - B_1\dot{x}_2 - K_{lt}x_2 - B_2(\dot{x}_2 - \dot{x}_1) - K_{se}(x_2 - x_1) = 0 . \quad (1.2)$$

Node 3:

$$F_{ant} + B_1\dot{x}_3 + K_{lt}x_3 - B_2(\dot{x}_4 - \dot{x}_3) - K_{se}(x_4 - x_3) = 0 . \quad (1.3)$$

Note that T_{ag} and T_{ant} are the tensions generated by agonist and antagonist muscles with designated direction as shown in Fig. 1.1, and

$$T_{ag} = B_2(\dot{x}_2 - \dot{x}_1) + K_{se}(x_2 - x_1)$$

$$T_{ant} = B_2(\dot{x}_4 - \dot{x}_3) + K_{se}(x_4 - x_3) .$$

(1.4)

Next, we let

$$J = \frac{J_p}{r} \times 5.2087 \times 10^3, \ B = \frac{B_p}{r} \times 5.2087 \times 10^3, \text{ and } K = \frac{K_p}{r} \times 5.2087 \times 10^3$$

and rewrite Eq. (1.1) as

$$B_2(\dot{x}_2 + \dot{x}_3 - \dot{x}_1 - \dot{x}_4) + K_{se} (x_2 + x_3 - x_1 - x_4) = J\ddot{x} + B\dot{x} + Kx .$$

(1.5)

We have assumed that there is an initial displacement from equilibrium at primary position for springs K_{lt} and K_{se} since the muscle is 3 mm longer than at equilibrium. That is

$$x_1 = x - x_{p1}$$
$$x_4 = x + x_{p4}.$$

To reduce the node equations to a single differential equation, we eliminate variables as before using the operating point analysis method. To this effect, we introduce the following variables and constants:

$$\hat{x} = x - x(0)$$
$$\hat{\theta} = \theta - \theta(0)$$
$$\hat{x}_1 = x_1 - x_1(0)$$
$$\hat{x}_2 = x_2 - x_2(0)$$
$$\hat{x}_3 = x_3 - x_3(0)$$
$$\hat{x}_4 = x_4 - x_4(0)$$
$$\hat{F}_{ag} = F_{ag} - F_{ag}(0)$$
$$\hat{F}_{ant} = F_{ant} - F_{ant}(0)$$
$$K_{st} = K_{se} + K_{lt}$$
$$B_{12} = B_1 + B_2 .$$

Note that with $\hat{x} = \hat{x}_1 = \hat{x}_4$, all derivative terms are zero at steady-state. Eliminating the derivative terms in Eqs. (1.2), (1.3), and (1.5), gives

$$F_{ag}(0) = K_{st}x_2(0) - K_{se}x_1(0)$$

$$F_{ant}(0) = K_{se}x_4(0) - K_{st}x_3(0)$$

(1.6)

$$K_{se} (x_2(0) + x_3(0) - x_1(0) - x_4(0)) = Kx(0) .$$

Subtracting the muscle node equations in Eq. (1.6) at steady-state for use later, gives

$$F_{ag}(0) - F_{ant}(0) = K_{st}(x_2(0) + x_3(0)) - K_{se}(x_1(0) + x_4(0)) . \tag{1.7}$$

Substituting the operating point variables and initial conditions into Eq. (1.2) gives

$$\hat{F}_{ag} + F_{ag}(0) - B_1\dot{\hat{x}}_2 - K_{lt}(\hat{x}_2 + x_2(0)) - B_2(\dot{\hat{x}}_2 - \dot{\hat{x}}_1) \\ - K_{se}(\hat{x}_2 + x_2(0) - \hat{x}_1 - x_1(0)) = 0 . \tag{1.8}$$

Removing the initial condition terms in Eq. (1.8) using steady-state terms from Eq. (1.6), yields

$$\hat{F}_{ag} = B_1\dot{\hat{x}}_2 + K_{lt}\hat{x}_2 + B_2(\dot{\hat{x}}_2 - \dot{\hat{x}}) + K_{se}(\hat{x}_2 - \hat{x}) \tag{1.9}$$

where $\hat{x}_1$ has been replaced by $\hat{x}$. After repeating this for Eqs. (1.3) and (1.5), we have

$$\hat{F}_{ant} = B_2(\dot{\hat{x}} - \dot{\hat{x}}_3) - B_1\dot{\hat{x}}_3 + K_{se}\hat{x} - K_{st}\hat{x}_3 \tag{1.10}$$

$$B_2(\dot{\hat{x}}_2 + \dot{\hat{x}}_3 - 2\dot{\hat{x}}) + K_{se}(\hat{x}_2 + \hat{x}_3 - 2\hat{x}) = J\ddot{\hat{x}} + B\dot{\hat{x}} + K\hat{x} \tag{1.11}$$

where $\hat{x}_4$ has been replaced by $\hat{x}$. We next apply the Laplace transform to Eqs. (1.9), (1.10), and (1.11), giving

$$\hat{F}_{ag}(s) = \hat{X}_2(s(B_1 + B_2) + K_{st}) - \hat{X}(sB_2 + K_{se}) \tag{1.12}$$

$$\hat{F}_{ant}(s) = \hat{X}(sB_2 + K_{se}) - \hat{X}_3(s(B_1 + B_2) + K_{st}) \tag{1.13}$$

$$(sB_2 + K_{se})\left(\hat{X}_2 + \hat{X}_3 - 2\hat{X}\right) = \hat{X}\left(Js^2 + Bs + K\right) . \tag{1.14}$$

Rearranging Eqs. (1.12) and (1.13) yields

$$\hat{X}_2 = \frac{\hat{F}_{ag}(s) + \hat{X}(sB_2 + K_{se})}{(sB_{12} + K_{st})}$$

$$\hat{X}_3 = \frac{(sB_2 + K_{se})\hat{X} - \hat{F}_{ant}(s)}{(sB_{12} + K_{st})}$$

and summing them together yields

$$\hat{X}_2 + \hat{X}_3 = \frac{\hat{F}_{ag}(s) + 2\hat{X}(sB_2 + K_{se}) - \hat{F}_{ant}(s)}{(sB_{12} + K_{st})} . \tag{1.15}$$

Substituting $\hat{X}_2 + \hat{X}_3$ (Eq. (1.15)) into Eq. (1.14) gives

$$(sB_2 + K_{se})\left(\frac{\hat{F}_{ag}(s) + 2\hat{X}(sB_2 + K_{se}) - \hat{F}_{ant}(s)}{(sB_{12} + K_{st})} - 2\hat{X}\right) = \hat{X}\left(Js^2 + Bs + K\right) . \tag{1.16}$$

After multiplying both sides of Eq. (1.16) by $(sB_{12} + K_{st})$, we have

$$(sB_2 + K_{se})\left(\hat{F}_{ag}(s) - \hat{F}_{ant}(s)\right) + 2\hat{X}(sB_2 + K_{se})^2 - 2\hat{X}(sB_2 + K_{se})(sB_{12} + K_{st})$$
$$= \hat{X}(sB_{12} + K_{st})\left(Js^2 + Bs + K\right) . \tag{1.17}$$

Simplifying Eq. (1.17) gives

$$(sB_2 + K_{se})\left(\hat{F}_{ag}(s) - \hat{F}_{ant}(s)\right) = \hat{X}\left(C_3s^3 + C_2s^2 + C_1s + C_0\right) \tag{1.18}$$

where

$$C_3 = JB_{12}$$
$$C_2 = JK_{st} + B_{12}B + 2B_1B_2$$
$$C_1 = 2B_1K_{se} + 2B_2K_{lt} + B_{12}K + K_{st}B$$
$$C_0 = K_{st}K + 2K_{lt}K_{se} .$$

We now transform back into time domain using the inverse Laplace transform, yielding

$$B_2\left(\dot{\hat{F}}_{ag} - \dot{\hat{F}}_{ant}\right) + K_{se}\left(\hat{F}_{ag} - \hat{F}_{ant}\right) = C_3\dddot{\hat{x}} + C_2\ddot{\hat{x}} + C_1\dot{\hat{x}} + C_0\hat{x} . \tag{1.19}$$

With $\dot{\hat{F}}_{ag} = \dot{F}_{ag}$, $\dot{\hat{F}}_{ant} = \dot{F}_{ant}$, $\hat{F}_{ag} = F_{ag} - F_{ag}(0)$ and $\hat{F}_{ant} = F_{ant} - F_{ant}(0)$, we have

$$\hat{F}_{ag} - \hat{F}_{ant} = F_{ag} - F_{ag}(0) - F_{ant} + F_{ant}(0) . \tag{1.20}$$

From Eq. (1.7), we have

$$F_{ag}(0) - F_{ant}(0) = K_{st}(x_2(0) + x_3(0)) - K_{se}(x_1(0) + x_4(0))$$

and with

$$x_1(0) = x(0) - x_{p1} \text{ and } x_4(0) = x(0) + x_{p4} = x(0) + x_{p1}$$

and by assuming identical muscles, we have

$$F_{ag}(0) - F_{ant}(0) = K_{st}(x_2(0) + x_3(0)) - 2K_{se}x(0) . \tag{1.21}$$

From Eq. (1.6), we have

$$K_{se}(x_2(0) + x_3(0) - x_1(0) - x_4(0)) = Kx(0)$$

and after removing $x_1(0)$ and $x_4(0)$ as before, gives

$$K_{se}(x_2(0) + x_3(0) - 2x(0)) = Kx(0)$$

and rearranging the previous equation, yields

$$x_2(0) + x_3(0) = \left(\frac{K}{K_{se}} + 2\right)x(0) .$$

When $x_2(0) + x_3(0)$ is substituted into Eq. (1.21), we have

$$F_{ag}(0) - F_{ant}(0) = \left(K_{st} \left(\frac{K}{K_{se}} + 2 \right) - 2K_{se} \right) x(0) \ . \tag{1.22}$$

With Eqs. (1.20) and (1.22) inserted into Eq. (1.19), we have

$$B_2 \left(\dot{F}_{ag} - \dot{F}_{ant} \right) + K_{se} \left(F_{ag} - F_{ant} \right) - K_{se} \left(F_{ag}(0) - F_{ant}(0) \right)$$
$$= C_3 \dddot{\hat{x}} + C_2 \ddot{\hat{x}} + C_1 \dot{\hat{x}} + C_0 \hat{x} \ .$$

Then with Eq. (1.22) $\left(F_{ag}(0) - F_{ant}(0) \right)$ substituted into the previous equation, we have

$$B_2 \left(\dot{F}_{ag} - \dot{F}_{ant} \right) + K_{se} \left(F_{ag} - F_{ant} \right) - K_{se} \left(K_{st} \left(\frac{K}{K_{se}} + 2 \right) - 2K_{se} \right) x(0)$$
$$= C_3 \dddot{\hat{x}} + C_2 \ddot{\hat{x}} + C_1 \dot{\hat{x}} + C_0 \left(\hat{x} - x(0) \right) \ . \tag{1.23}$$

To reduce Eq. (1.23) further, we note that

$$K_{se} \left(K_{st} \left(\frac{K}{K_{se}} + 2 \right) - 2K_{se} \right) x(0) = C_0 x(0) = (K_{st} K + 2K_{lt} K_{se}) x(0) \tag{1.24}$$

or

$$K_{st} K + 2K_{st} - 2K_{se} = K_{st} K + 2K_{lt} K_{se} \ .$$

Since the left-hand side of Eq. (1.24) equals the right-hand side of Eq. (1.24), Eq. (1.23) becomes

$$B_2 \left(\dot{F}_{ag} - \dot{F}_{ant} \right) + K_{se} \left(F_{ag} - F_{ant} \right) = C_3 \dddot{\hat{x}} + C_2 \ddot{\hat{x}} + C_1 \dot{\hat{x}} + C_0 \hat{x} \ . \tag{1.25}$$

In Eq. (1.25), we exchange x for θ, $\left(x = \frac{\theta}{5208.7} \right)$, giving

$$\delta \left(B_2 \left(\dot{F}_{ag} - \dot{F}_{ant} \right) + K_{se} \left(F_{ag} - F_{ant} \right) \right) = \dddot{\theta} + P_2 \ddot{\theta} + P_1 \dot{\theta} + P_0 \theta \tag{1.26}$$

where

$$\delta = \frac{5208.7}{J B_{12}}$$

$$P_2 = \frac{J K_{st} + B_{12} B + 2B_1 B_2}{J B_{12}}$$

$$P_1 = \frac{2B_1 K_{se} + 2B_2 K_{lt} + B_{12} K + K_{st} B}{J B_{12}}$$

$$P_0 = \frac{K_{st} K + 2K_{lt} K_{se}}{J B_{12}} \ .$$

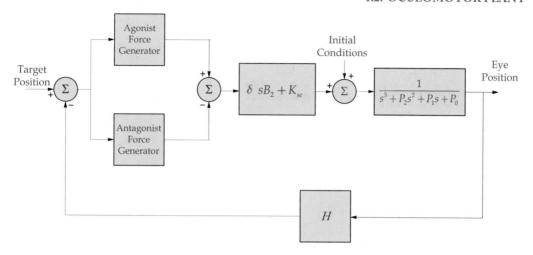

Figure 1.3: Block diagram of the 2009 linear homeomorphic saccadic eye movement model.

A block diagram for the system in Fig. 1.1 is shown in Fig. 1.3. The section of the diagram in the forward path is the 2009 linear homeomorphic saccadic eye movement model. The feedback element H is unity and is operational only when the eye is stable. It is clear that this model has a characteristic equation that is 3rd order rather than 4th order in the previous version of the model. This model is easier to use than the 1995 model without sacrificing accuracy.

1.2.2 NEURAL INPUT

In Section 3.3 of Book 1, we modeled the neural input to the saccade system as a pulse-step waveform. This input has been used in many studies because of its simplicity and ease of use (Bahill et al., 1980; Enderle and Wolfe, 1988; Enderle et al., 1984). To create a more realistic input based on physiological evidence, a pulse-slide-step input is used as shown in Fig. 1.4 (based on Goldstein, H., 1983). The slide is an exponential transition from the pulse to the step. This model is consistent with the data published in the literature (for example, see Fig. 4 in Robinson, D., 1981 and Fig. 2 in Gisbergen et al., 1981). The diagram in Fig. 1.4 (Top) closely approximates the data shown in Fig. 1.4 (Bottom) for the agonist input. An explanation of the neural input will be given in a later section.

At steady-state, the eye is held steady by the agonist and antagonist inputs F_{g0} and F_{t0}. We typically define the time when the target moves as $t = 0$. This is a common assumption since many simulation studies ignore the latent period and focus on the actual movement (see the time axis in Fig. 1.4, Top and Bottom).

The overall agonist pulse occurs in the interval $0 - T_2$, with a more complex behavior than the pulse described in Section 3.3 of Book 1. We view the overall pulse process as the intention of the

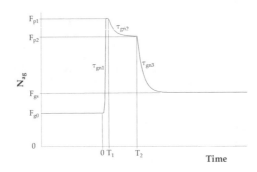

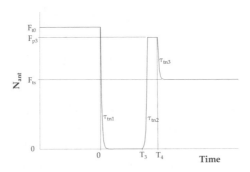

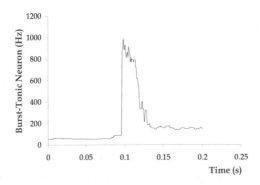

Figure 1.4: Neural input to the horizontal saccade system. (Top) Agonist input. (Middle) Antagonist input. (Bottom) Discharge rate of a single burst-tonic neuron during a saccade (agonist input). Details of the experiment and training for (Bottom) are reported elsewhere [Sparks et al., 1976. Size and distribution of movement fields in the monkey superior colliculus, *Brain Research*, vol. 113, pp. 21–34.]. (Data provided personally by Dr. David Sparks.)

system, which is limited by its physical capabilities. The start of the pulse occurs with an exponential rise from the initial firing rate, F_{g0}, to peak magnitude, F_{p1}, with a time constant τ_{gn1}. At T_1, the input decays to F_{p2}, with a time constant τ_{gn2}. The slide occurs at T_2, with a time constant τ_{gn3}, to F_{gs}, the force necessary to hold the eye at its destination. The input F_{gs} is applied during the step portion of the input.

At $t = 0$, the antagonist neural input is completely inhibited and exponentially decays to zero from F_{t0} with time constant τ_{tn1}. At time T_3, the antagonist input exponentially increases with time constant τ_{tn2}. The antagonist neural input shown in Fig. 1.4 (Middle) includes a PIRB pulse with duration of $T_4 - T_3$. At T_4, the antagonist input exponentially decays to F_{ts}, with a time constant τ_{tn3}. If no PIRB occurs in the antagonist input, the input exponentially rises to F_{ts} with time constant τ_{tn2}.

The agonist pulse includes an interval (T_1) that is constant for saccades of all sizes as supported by physiological evidence (Enderle, J., 2002; Zhou et al., 2009). We choose to model the change in the firing rate with an exponential function as this seems to match the data fairly well.

After complete inhibition, the antagonist neural input has a brief excitatory pulse starting at T_3 with duration of approximately 10 ms. Enderle proposed that this burst is generated by PIRB, a property which contributes to the post-saccade phenomena such as dynamic and glissadic overshoot (Enderle, J., 2002).

Based on the diagram in Fig. 1.4 and assuming the exponential terms reach steady-state at $t = 5\tau$, the equations for N_{ag} and N_{ant} are written as the following:

$$
\begin{aligned}
N_{ag} = \ &F_{g0}u(-t) \\
&+ \left(F_{g0} + (F_{p1} - F_{g0})e^{\frac{t}{\tau_{gn1}} - 5} \right) \left(u(t) - u\left(t - 5\tau_{gn1} \right) \right) \\
&+ F_{p1} \left(u\left(t - 5\tau_{gn1} \right) - u\left(t - T_1 \right) \right) \\
&+ \left(F_{p2} + (F_{p1} - F_{p2})e^{\frac{(T_1 - t)}{\tau_{gn2}}} \right) \left(u\left(t - T_1 \right) - u\left(t - T_2 \right) \right) \\
&+ \left(F_{gs} + \left(F_{p2} + \left(F_{p1} - F_{p2} \right) e^{\frac{(T_1 - T_2)}{\tau_{gn2}}} - F_{gs} \right) e^{\frac{(T_2 - t)}{\tau_{gn3}}} \right) u\left(t - T_2 \right)
\end{aligned}
\tag{1.27}
$$

and

$$
\begin{aligned}
N_{ant} = \ &F_{t0}u(-t) \\
&+ F_{t0}e^{\frac{-t}{\tau_{tn1}}} \left(u(t) - u\left(t - T_3 \right) \right) \\
&+ \left(F_{t0}e^{\frac{-T_3}{\tau_{tn1}}} + \left(F_{p3} - F_{t0}e^{\frac{-T_3}{\tau_{tn1}}} \right) e^{\frac{(t - T_3)}{\tau_{tn2}} - 5} \right) \left(u\left(t - T_3 \right) - u\left(t - T_3 - 5\tau_{tn2} \right) \right) \\
&+ F_{p3} \left(u\left(t - T_3 - 5\tau_{tn2} \right) - u\left(t - T_4 \right) \right) \\
&+ \left(F_{ts} + \left(F_{p3} - F_{ts} \right) e^{\frac{(T_4 - t)}{\tau_{tn3}}} \right) u\left(t - T_4 \right).
\end{aligned}
\tag{1.28}
$$

Note that Eqs. (1.27) and (1.28) are written in terms of intervals. Further, we assume that $5\tau_{gn1} < T_1$ and $T_3 + 5\tau_{tn2} < T_4$, which simplifies analysis.

As before, the agonist and antagonist active-state tensions are defined as low-pass filtered neural inputs:

$$\dot{F}_{ag} = \frac{N_{ag} - F_{ag}}{\tau_{ag}} \tag{1.29}$$

$$\dot{F}_{ant} = \frac{N_{ant} - F_{ant}}{\tau_{ant}} \tag{1.30}$$

where

$$\tau_{ag} = \tau_{gac} \left(u(t - T_1) - u\left(t - T_2\right) \right) + \tau_{gde} u\left(t - T_2\right) \tag{1.31}$$

$$\tau_{ant} = \tau_{tde} \left(u(t) - u\left(t - T_3\right) \right) + \tau_{tac} \left(u\left(t - T_3\right) - u\left(t - T_4\right) \right) + \tau_{tde} u\left(t - T_4\right) . \tag{1.32}$$

The activation and deactivation time constants represent the different dynamic characteristics of muscle under increasing and decreasing stimulation. Shown in Fig. 1.5 are the agonist and antagonist active-state tensions, derived from low-pass filtering the neural inputs.

The analytical solutions for F_{ag} and F_{ant} are derived from Eqs. (1.29) and (1.30) using the neural inputs from Eqs. (1.27) and (1.28). We use following constants for ease in solution:

$$A = F_{p1} - F_{g0}$$
$$B = F_{p1} - F_{p2}$$
$$C = F_{p2} + \left(F_{p1} - F_{p2}\right) e^{\frac{(T_1 - T_2)}{\tau_{gn2}}} - F_{gs}$$
$$D = F_{t0} e^{\frac{-T_3}{\tau_{tn1}}}$$
$$E = F_{p3} - F_{t0} e^{\frac{-T_3}{\tau_{tn1}}}$$
$$F = F_{p3} - F_{ts} .$$

F_{ag} *Response*   The agonist *natural response* has one time constant, τ_{ag}, and, in general, has a response of $F_{ag_n} = K_g e^{-\frac{t}{\tau_{ag}}}$. Since τ_{ag} is defined in two intervals according to Eq. (1.31), we need to be careful in how this is handled. There are four inputs applied to the system, and so we use superposition to write the solution.

$$
\begin{aligned}
F_{ag_n}(t) = {} & K_{g0} e^{\frac{-t}{\tau_{gac}}} \left(u(t) - u\left(t - 5\tau_{gn1}\right) \right) \\
& + K_{g1} e^{\frac{(5\tau_{gn1} - t)}{\tau_{gac}}} \left(u\left(t - 5\tau_{gn1}\right) - u\left(t - T_1\right) \right) \\
& + K_{g2} e^{\frac{(T_1 - t)}{\tau_{gac}}} \left(u\left(t - T_1\right) - u\left(t - T_2\right) \right) \\
& + K_{g3} e^{\frac{(T_2 - t)}{\tau_{gde}}} u\left(t - T_2\right)
\end{aligned}
\tag{1.33}
$$

where K_{g0}, K_{g1}, K_{g2}, and K_{g3} are unknown constants to be determined from the initial conditions.

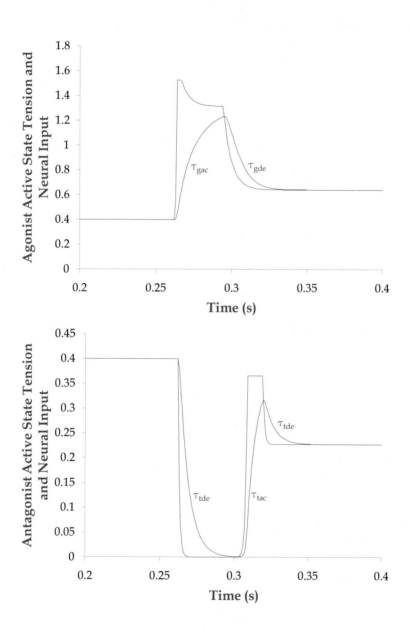

Figure 1.5: (Top) Agonist active state tension (red) and neural input (blue). (Bottom) Antagonist active state tension (red) and neural input (blue).

The agonist *forced response* is based on the form of the input. We substitute the input for each interval in Eq. (1.27) into Eq. (1.29), and find, after some work, that

$$
\begin{aligned}
F_{ag_f}(t) = {} & F_{g0}u(-t) \\
& + \left(F_{g0} + L_{g1}e^{\frac{t}{\tau_{gn1}}-5} \right) \left(u(t) - u\left(t - 5\tau_{gn1}\right) \right) \\
& + F_{p1}\left(u\left(t - 5\tau_{gn1}\right) - u\left(t - T_1\right) \right) \\
& + \left(F_{p2} + L_{g2}e^{\frac{(T_1-t)}{\tau_{gn2}}} \right) \left(u\left(t - T_1\right) - u\left(t - T_2\right) \right) \\
& + \left(F_{gs} + L_{g3}e^{\frac{(T_2-t)}{\tau_{gn3}}} \right) u\left(t - T_2\right)
\end{aligned}
\tag{1.34}
$$

where

$$
\begin{aligned}
L_{g1} &= \frac{\tau_{gn1}}{\tau_{gac} + \tau_{gn1}} A \\
L_{g2} &= \frac{\tau_{gn2}}{\tau_{gn2} - \tau_{gac}} B \\
L_{g3} &= \frac{\tau_{gn3}}{\tau_{gn3} - \tau_{gde}} C \; .
\end{aligned}
$$

The total agonist active state tension equals $F_{ag} = F_{ag_n} + F_{ag_f}$ and is given by

$$
\begin{aligned}
F_{ag}(t) = {} & K_{g0}e^{\frac{-t}{\tau_{gac}}} \left(u(t) - u\left(t - 5\tau_{gn1}\right) \right) + K_{g1}e^{\frac{(5\tau_{gn1}-t)}{\tau_{gac}}} \left(u\left(t - 5\tau_{gn1}\right) - u\left(t - T_1\right) \right) \\
& + K_{g2}e^{\frac{(T_1-t)}{\tau_{gac}}} \left(u\left(t - T_1\right) - u\left(t - T_2\right) \right) + K_{g3}e^{\frac{(T_2-t)}{\tau_{gde}}} u(t - T_2) + F_{g0}u(-t) \\
& + \left(F_{g0} + L_{g1}e^{\frac{t}{\tau_{gn1}}-5} \right) \left(u(t) - u\left(t - 5\tau_{gn1}\right) \right) + F_{p1}\left(u(t - 5\tau_{gn1}) - u\left(t - T_1\right) \right) \\
& + \left(F_{p2} + L_{g2}e^{\frac{(T_1-t)}{\tau_{gn2}}} \right) \left(u\left(t - T_1\right) - u\left(t - T_2\right) \right) + \left(F_{gs} + L_{g3}e^{\frac{(T_2-t)}{\tau_{gn3}}} \right) u\left(t - T_2\right) \\
= {} & F_{g0}u(-t) + \left(K_{g0}e^{\frac{-t}{\tau_{gac}}} + F_{g0} + L_{g1}e^{\frac{t}{\tau_{gn1}}-5} \right) \left(u(t) - u\left(t - 5\tau_{gn1}\right) \right) \\
& + \left(K_{g1}e^{\frac{(5\tau_{gn1}-t)}{\tau_{gac}}} + F_{p1} \right) \left(u\left(t - 5\tau_{gn1}\right) - u\left(t - T_1\right) \right) \\
& + \left(K_{g2}e^{\frac{(T_1-t)}{\tau_{gac}}} + F_{p2} + L_{g2}e^{\frac{(T_1-t)}{\tau_{gn2}}} \right) \left(u\left(t - T_1\right) - u\left(t - T_2\right) \right) \\
& + \left(K_{g3}e^{\frac{(T_2-t)}{\tau_{gde}}} + F_{gs} + L_{g3}e^{\frac{(T_2-t)}{\tau_{gn3}}} \right) u\left(t - T_2\right) \; .
\end{aligned}
\tag{1.35}
$$

To evaluate unknown constants, K_{g0}, K_{g1}, K_{g2}, and K_{g3}, in Eq. (1.35), we use the initial condition and the fact that F_{ag} is continuous in time. For example, at switch time T_1, $F_{ag}\left(T_1^-\right) = F_{ag}\left(T_1^+\right)$. Thus,

$t = 0:$ $F_{ag} = F_{g0}$

$$F_{g0} = K_{g0} + F_{g0} + L_{g1}e^{-5} \Rightarrow K_{g0} = F_{g0} - \left(F_{g0} + L_{g1}e^{-5}\right)$$

$t = 5\tau_{gn1}:$ F_{ag} is continuous

$$K_{g0}e^{-\frac{5\tau_{gn1}}{\tau_{gac}}} + F_{g0} + L_{g1} = K_{g1} + F_{p1} \Rightarrow K_{g1} = K_{g0}e^{-\frac{5\tau_{gn1}}{\tau_{gac}}} + F_{g0} + L_{g1} - F_{p1}$$

$t = T_1:$ F_{ag} is continuous

$$K_{g1}e^{-\frac{T_1-5\tau_{gn1}}{\tau_{gac}}} + F_{p1} = K_{g2} + F_{p2} + L_{g2} \Rightarrow K_{g2} = K_{g1}e^{-\frac{T_1-5\tau_{gn1}}{\tau_{gac}}} + F_{p1} - \left(F_{p2} + L_{g2}\right)$$

$t = T_2:$ F_{ag} is continuous

$$K_{g2}e^{\frac{(T_1-T_2)}{\tau_{gac}}} + F_{p2} + L_{g2}e^{\frac{(T_1-T_2)}{\tau_{gn2}}} = K_{g3} + F_{gs} + L_{g3}$$
$$\Rightarrow K_{g3} = K_{g2}e^{\frac{(T_1-T_2)}{\tau_{gac}}} + F_{p2} + L_{g2}e^{\frac{(T_1-T_2)}{\tau_{gn2}}} - \left(F_{gs} + L_{g3}\right).$$

Based on these calculations and simplifying Eq. (1.35) by bringing together the unit step functions as single terms, F_{ag} is

$$\begin{aligned}
F_{ag} = {}& F_{g0}u(-t) \\
& + \left(K_{g0}e^{\frac{-t}{\tau_{gac}}} + F_{g0} + L_{g1}e^{\frac{t}{\tau_{gn1}}-5}\right)u(t) \\
& + \left(M_{g1}e^{\frac{(5\tau_{gn1}-t)}{\tau_{gac}}} + F_{p1} - F_{g0} - L_{g1}e^{\frac{t}{\tau_{gn1}}-5}\right)u\left(t - 5\tau_{gn1}\right) \\
& + \left(M_{g2}e^{\frac{(T_1-t)}{\tau_{gac}}} + F_{p2} + L_{g2}e^{\frac{(T_1-t)}{\tau_{gn2}}} - F_{p1}\right)u\left(t - T_1\right) \\
& + \left(M_{g3}e^{\frac{(T_2-t)}{\tau_{gac}}} + K_{g3}e^{\frac{(T_2-t)}{\tau_{gde}}} + F_{gs} + L_{g3}e^{\frac{(T_2-t)}{\tau_{gn3}}} - F_{p2} - L_{g2}e^{\frac{(T_1-t)}{\tau_{gn2}}}\right)u\left(t - T_2\right)
\end{aligned} \tag{1.36}$$

where

$$M_{g1} = K_{g1} - K_{g0}e^{-\frac{5\tau_{gn1}}{\tau_{gac}}}$$
$$M_{g2} = K_{g2} - K_{g1}e^{\frac{5\tau_{gn1}-T_1}{\tau_{gac}}}$$
$$M_{g3} = -K_{g2}e^{\frac{(T_1-T_2)}{\tau_{gac}}}.$$

F_{ant} *Response*   The antagonist *natural response* has one time constant, τ_{ant}, that is defined in two intervals as in Eq. (1.32). As before with several inputs applied, we use superposition to write the solution.

$$
\begin{aligned}
F_{ant_n} = {} & K_{t0}e^{-\frac{t}{\tau_{tde}}}\left(u(t) - u\left(t - T_3\right)\right) \\
& + K_{t1}e^{-\frac{t-T_3}{\tau_{tac}}}\left(u\left(t - T_3\right) - u\left(t - T_3 - 5\tau_{tn2}\right)\right) \\
& + K_{t2}e^{-\frac{t-T_3-5\tau_{tn2}}{\tau_{tac}}}\left(u\left(t - T_3 - 5\tau_{tn2}\right) - u\left(t - T_4\right)\right) \\
& + K_{t3}e^{-\frac{t-T_4}{\tau_{tde}}}u\left(t - T_4\right)
\end{aligned}
\tag{1.37}
$$

where $K_{t0}, K_{t1}, K_{t2},$ and K_{t3} are unknown constants to be determined from the initial conditions.

The antagonist *forced response* is based on the form of the input. We substitute the input for each interval in Eq. (1.28) into Eq. (1.30), and have

$$
\begin{aligned}
F_{ant_f}(t) = {} & F_{t0}u(-t) \\
& + L_{t1}e^{\frac{-t}{\tau_{tn1}}}\left(u(t) - u\left(t - T_3\right)\right) \\
& + \left(D + L_{t2}e^{\frac{t-T_3}{\tau_{tn2}} - 5}\right)\left(u\left(t - T_3\right) - u\left(t - T_3 - 5\tau_{tn2}\right)\right) \\
& + F_{p3}\left(u\left(t - T_3 - 5\tau_{tn2}\right) - u\left(t - T_4\right)\right) \\
& + \left(F_{ts} + L_{t3}e^{\frac{(T_4-t)}{\tau_{tn3}}}\right)u\left(t - T_4\right)
\end{aligned}
\tag{1.38}
$$

where

$$
L_{t1} = \frac{\tau_{tn1}}{\tau_{tn1} - \tau_{tde}}F_{t0}
$$

$$
L_{t2} = \frac{\tau_{tn2}}{\tau_{tn2} + \tau_{tac}}E
$$

$$
L_{t3} = \frac{\tau_{tn3}}{\tau_{tn3} - \tau_{tde}}F .
$$

The total antagonist active state tension equals $F_{ant} = F_{ant_n} + F_{ant_f}$ and is given by

$$
\begin{aligned}
F_{ant}(t) = {} & K_{t0}e^{\frac{-t}{\tau_{tde}}}\left(u(t) - u\left(t - T_3\right)\right) + K_{t1}e^{\frac{(T_3-t)}{\tau_{tac}}}\left(u\left(t - T_3\right) - u\left(t - T_3 - 5\tau_{tn2}\right)\right) \\
& + K_{t2}e^{\frac{T_3+5\tau_{tn2}-t}{\tau_{tac}}}\left(u\left(t - T_3 - 5\tau_{tn2}\right) - u\left(t - T_4\right)\right) + K_{t3}e^{\frac{(T_4-t)}{\tau_{tde}}}u\left(t - T_4\right) + F_{t0}u(-t) \\
& + L_{t1}e^{\frac{-t}{\tau_{tn1}}}\left(u(t) - u\left(t - T_3\right)\right) + \left(D + L_{t2}e^{\frac{t-T_3}{\tau_{tn2}} - 5}\right)\left(u\left(t - T_3\right) - u\left(t - T_3 - 5\tau_{tn2}\right)\right) \\
& + F_{p3}\left(u\left(t - T_3 - 5\tau_{tn2}\right) - u\left(t - T_4\right)\right) + \left(F_{ts} + L_{t3}e^{\frac{(T_4-t)}{\tau_{tn3}}}\right)u\left(t - T_4\right) \\
= {} & F_{t0}u(-t) + \left(K_{t0}e^{\frac{-t}{\tau_{tde}}} + L_{t1}e^{\frac{-t}{\tau_{tn1}}}\right)\left(u(t) - u\left(t - T_3\right)\right) \\
& + \left(K_{t1}e^{\frac{(T_3-t)}{\tau_{tac}}} + D + L_{t2}e^{\frac{t-T_3}{\tau_{tn2}} - 5}\right)\left(u\left(t - T_3\right) - u\left(t - T_3 - 5\tau_{tn2}\right)\right) \\
& + \left(K_{t2}e^{\frac{(T_3+5\tau_{tn2}-t)}{\tau_{tac}}} + F_{p3}\right)\left(u\left(t - T_3 - 5\tau_{tn2}\right) - u\left(t - T_4\right)\right) \\
& + \left(K_{t3}e^{\frac{(T_4-t)}{\tau_{tde}}} + F_{ts} + L_{t3}e^{\frac{(T_4-t)}{\tau_{tn3}}}\right)u\left(t - T_4\right) .
\end{aligned}
\tag{1.39}
$$

To evaluate unknown constants, K_{t0}, K_{t1}, K_{t2}, and K_{t3}, in Eq. (1.39), we use the initial condition and the fact that F_{ant} is continuous in time. For example, at switch time T_3, $F_{ant}\left(T_3^-\right) = F_{ant}\left(T_3^+\right)$. Thus,

$t = 0:$ $F_{ant} = F_{t0}$

$$F_{t0} = K_{t0} + L_{t1} \Rightarrow K_{t0} = F_{t0} - L_{t1}$$

$t = T_3:$ F_{ant} is continuous

$$L_{t1}e^{\frac{-T_3}{\tau_{tn1}}} + K_{t0}e^{\frac{T_3}{\tau_{tde}}} = K_{t1} + D + L_{t2}e^{-5} \Rightarrow K_{t1} = L_{t1}e^{\frac{T_3}{\tau_{tn1}}} + K_{t0}e^{\frac{T_3}{\tau_{tde}}} - \left(D + L_{t2}e^{-5}\right)$$

$t = T_3 + 5\tau_{tn2}:$ F_{ant} is continuous

$$K_{t1}e^{\frac{-5\tau_{tn2}}{\tau_{tac}}} + D + L_{t2} = K_{t2} + F_{p3} \Rightarrow K_{t2} = K_{t1}e^{\frac{-5\tau_{tn2}}{\tau_{tac}}} + D + L_{t2} - F_{p3}$$

$t = T_4:$ F_{ant} is continuous

$$K_{t2}e^{\frac{(T_3+5\tau_{tn2}-T_4)}{\tau_{tac}}} + F_{p3} = K_{t3} + F_{ts} + L_{t3} \Rightarrow K_{t3} = K_{t2}e^{\frac{(T_3+5\tau_{tn2}-T_4)}{\tau_{tac}}} + F_{p3} - (F_{ts} + L_{t3}) \ .$$

Based on these calculations and simplifying Eq. (1.38) by bringing together the unit step functions as single terms, F_{ant} is

$$
\begin{aligned}
F_{ant}(t) = {} & F_{t0}u(-t) \\
& + \left(K_{t0}e^{-\frac{t}{\tau_{tde}}} + L_{t1}e^{-\frac{t}{\tau_{tn1}}}\right)u(t) \\
& + \left(K_{t1}e^{\frac{(T_3-t)}{\tau_{tac}}} - M_{t1}e^{\frac{(T_3-t)}{\tau_{tde}}} + D + L_{t2}e^{\frac{(t-T_3)}{\tau_{tn2}}-5} - L_{t1}e^{\frac{-t}{\tau_{tn1}}}\right)u\left(t - T_3\right) \qquad (1.40) \\
& + \left(M_{t2}e^{\frac{(T_3+5\tau_{tn2}-t)}{\tau_{tac}}} + F_{p3} - D - L_{t2}e^{\frac{(t-T_3)}{\tau_{tn2}}-5}\right)u\left(t - T_3 - 5\tau_{tn2}\right) \\
& + \left(M_{t3}e^{\frac{(T_4-t)}{\tau_{tac}}} + K_{t3}e^{\frac{(T_4-t)}{\tau_{tde}}} - F + L_{t3}e^{\frac{(T_4-t)}{\tau_{tn3}}}\right)u\left(t - T_4\right)
\end{aligned}
$$

where

$$M_{t1} = K_{t0}e^{\frac{-T_3}{\tau_{tde}}}$$
$$M_{t2} = K_{t2} - K_{t1}e^{\frac{-5\tau_{tn2}}{\tau_{tac}}}$$
$$M_{t3} = -K_{t2}e^{\frac{(T_3+5\tau_{tn2}-T_4)}{\tau_{tac}}} \ .$$

The derivatives of active state tensions are found by using Eqs. (1.29) and (1.30). F_{g0}, F_{gs}, F_{t0} and F_{ts} are steady-state tensions determined from the following:

$$F = \begin{cases} 0.4 + 0.0175\,|\theta| & N \quad \text{for} \quad \theta \geq 0° \\ 0.4 - 0.0125\,|\theta| & N \quad \text{for} \quad \theta < 0° \ . \end{cases}$$

In practice, a short time delay of 1 ms is introduced to reflect the time it takes to send the signal from the Abducens and Oculomotor Nucleus to the muscles.

1.2.3   SACCADE RESPONSE

At this time, we wish to solve for the complete response for a saccade. To begin, the system given by Eq. (1.26) is repeated here for convenience as

$$\delta\left(B_2\left(\dot{F}_{ag} - \dot{F}_{ant}\right) + K_{se}\left(F_{ag} - F_{ant}\right)\right) = \dddot{\theta} + P_2\ddot{\theta} + P_1\dot{\theta} + P_0\theta$$

which is driven by the inputs F_{ag} and F_{ant} described by Eqs. (1.36) and (1.40). As usual, the solution is composed of the natural and forced response. We begin with the forced response and follow with the natural response.

Forced Response   To evaluate the forced response for Eq. (1.26), Eqs. (1.35), (1.39), and derivative terms are substituted into the left-hand side of Eq. (1.26). The forced response is a linear combination of unit step functions with exponential components. The unit step functions are $u(t)$, $u\left(t - 5\tau_{gn1}\right)$, $u\left(t - T_1\right)$, $u\left(t - T_2\right)$, $u\left(t - T_3\right)$, $u\left(t - T_3 - 5\tau_{tn2}\right)$, and $u\left(t - T_4\right)$. The exponential components are $e^{\frac{(t_n - t)}{\tau_{gac}}}$, $e^{\frac{(t_n - t)}{\tau_{gde}}}$, $e^{\frac{(t_n - t)}{\tau_{gn1}}}$, $e^{\frac{(t_n - t)}{\tau_{gn2}}}$, $e^{\frac{(t_n - t)}{\tau_{gn3}}}$, $e^{\frac{(t_n - t)}{\tau_{tac}}}$, $e^{\frac{(t_n - t)}{\tau_{tde}}}$, $e^{\frac{(t_n - t)}{\tau_{tn1}}}$, $e^{\frac{(t_n - t)}{\tau_{tn2}}}$, and $e^{\frac{(t_n - t)}{\tau_{tn3}}}$; the variable t_n represents the switch times. For example, we can write the left-hand side of Eq. (1.26) as

$$LHS = \delta \begin{bmatrix} \left(C_{1,1} + C_{1,2}e^{\frac{-t}{\tau_{gac}}} + C_{1,3}e^{\frac{-t}{\tau_{gde}}} + \cdots + C_{1,11}e^{\frac{-t}{\tau_{tn3}}}\right)u(t) \\ + \left(C_{2,1} + C_{2,2}e^{\frac{(5\tau_{gn1}-t)}{\tau_{gac}}} + C_{2,3}e^{\frac{(5\tau_{gn1}-t)}{\tau_{gde}}} + \cdots + C_{2,11}e^{\frac{(5\tau_{gn1}-t)}{\tau_{tn3}}}\right)u\left(t - 5\tau_{gn1}\right) + \\ \vdots \\ \left(C_{7,1} + C_{7,2}e^{\frac{(T_4-t)}{\tau_{gac}}} + C_{7,3}e^{\frac{(T_4-t)}{\tau_{gde}}} + \cdots + C_{7,11}e^{\frac{(T_4-t)}{\tau_{tn3}}}\right)u\left(t - T_4\right) \end{bmatrix}$$

$$(1.41)$$

where $C_{m,n}$ are the coefficients for each term whose values are listed in Table 1.1.

The forced response is

$$\theta_f(t) = \delta \begin{pmatrix} \left(A_{1,1} + A_{1,2}e^{-t/\tau_{gac}} + A_{1,3}e^{-t/\tau_{gde}} + \cdots + A_{1,11}e^{-t/\tau_{tn3}}\right)u(t) \\ + \begin{pmatrix} A_{2,1} + A_{2,2}e^{-(t-5\tau_{gn1})/\tau_{gac}} + A_{2,3}e^{-(t-5\tau_{gn1})/\tau_{gde}} + \cdots \\ + A_{2,11}e^{-(t-5\tau_{gn1})/\tau_{tn3}} \end{pmatrix}u\left(t - 5\tau_{gn1}\right) \\ \vdots \\ + \begin{pmatrix} A_{7,1} + A_{7,2}e^{-(t-T_4)/\tau_{gac}} + A_{7,3}e^{-(t-T_4)/\tau_{gde}} + \cdots \\ + A_{7,11}e^{-(t-T_4)/\tau_{tn3}} \end{pmatrix}u\left(t - T_4\right) \end{pmatrix}$$

$$(1.42)$$

Table 1.1: Coefficients of the terms in Eq. (1.41) (i.e., $C_{1,1} = K_{se}F_{g0}$, $C_{1,2} = \left(K_{se} - \frac{B_2}{\tau_{gac}}\right)K_{g0}$, $C_{2,1} = K_{se}\left(F_{p1} - F_{g0}\right)$, etc.) (continues).

LHS $= \delta\sum$	1	$e^{\frac{t_n-t}{\tau_{gac}}}$	$e^{\frac{t_n-t}{\tau_{gde}}}$	$e^{\frac{t_n-t}{\tau_{gn1}}}$	$e^{\frac{t_n-t}{\tau_{gn2}}}$	$e^{\frac{t_n-t}{\tau_{gn3}}}$
$u(t)$	$K_{se}F_{g0}$	$\left(K_{se} - \frac{B_2}{\tau_{gac}}\right)K_{g0}$	0	0	0	0
$u\left(t - 5\tau_{gn1}\right)$	$K_{se}\left(F_{p1} - F_{g0}\right)$	$\left(K_{se} - \frac{B_2}{\tau_{gac}}\right)M_{g1}$	0	$-\left(K_{se} + \frac{B_2}{\tau_{gn1}}\right)L_{g1}e^{-5}$	0	0
$u\left(t - \tau_1\right)$	$-K_{se}B$	$\left(K_{se} - \frac{B_2}{\tau_{gac}}\right)M_{g2}$	0	0	$\left(K_{se} - \frac{B_2}{\tau_{gn2}}\right)L_{g2}$	0
$u\left(t - \tau_2\right)$	$K_{se}\left(F_{gs} - F_{p2}\right)$	$\left(K_{se} - \frac{B_2}{\tau_{gac}}\right)M_{g3}$	$\left(K_{se} - \frac{B_2}{\tau_{gde}}\right)K_{g3}$	0	$-\left(K_{se} - \frac{B_2}{\tau_{gn2}}\right)L_{g2}e^{\frac{t_2-t_1}{\tau_{gn2}}}$	0
$u\left(t - \tau_3\right)$	$-K_{se}D$	0	0	0	0	0
$u\left(t - \tau_3\right) - 5\tau_{tn2}$	$-K_{se}\left(F_{p3} - D\right)$	0	0	0	0	$\left(K_{se} - \frac{B_2}{\tau_{gn3}}\right)L_{g3}$
$u\left(t - \tau_4\right)$	$K_{se}F$	0	0	0	0	0

Table 1.2: (continued) Coefficients of the terms in Eq. (1.41).

LHS $= \delta\sum$	$e^{\frac{t_n-t}{\tau_{tac}}}$	$e^{\frac{t_n-t}{\tau_{tde}}}$	$e^{\frac{t_n-t}{\tau_{tn1}}}$	$e^{\frac{t_n-t}{\tau_{tn2}}}$	$e^{\frac{t_n-t}{\tau_{tn3}}}$
$u(t)$	0	$-\left(K_{se}-\frac{B_2}{\tau_{tde}}\right)K_{t0}$	$-\left(K_{se}-\frac{B_2}{\tau_{tn1}}\right)L_{t1}$	0	0
$u(t-5\tau_{gn1})$	0	0	0	0	0
$u(t-\tau_1)$	0	0	0	0	0
$u(t-\tau_2)$	0	0	0	0	0
$u(t-\tau_3)$	$-\left(K_{se}-\frac{B_2}{\tau_{tac}}\right)K_{t1}$	$\left(K_{se}-\frac{B_2}{\tau_{tde}}\right)M_{t1}$	$\left(K_{se}-\frac{B_2}{\tau_{tn1}}\right)L_{t1}e^{-\frac{t_3}{\tau_{tn1}}}$	$-\left(K_{se}+\frac{B_2}{\tau_{tn2}}\right)L_{t2}e^{-5}$	0
$u(t-\tau_3-5\tau_{tn2})$	$-\left(K_{se}-\frac{B_2}{\tau_{tac}}\right)M_{t2}$	0	0	$\left(K_{se}+\frac{B_2}{\tau_{tn2}}\right)L_{t2}$	0
$u(t-\tau_4)$	$-\left(K_{se}-\frac{B_2}{\tau_{tac}}\right)M_{t3}$	$-\left(K_{se}-\frac{B_2}{\tau_{tde}}\right)K_{t3}$	0	0	$-\left(K_{se}-\frac{B_2}{\tau_{tn3}}\right)L_{t3}$

where the coefficients $A_{m,n}$ are functions of $C_{m,n}$ as shown in Table 1.1 by considering the right-hand side of Eq. (1.26). For example, $A_{1,n(n=1...11)} = \frac{\delta C_{1,n}}{D_n}$, where D_n is

$$D_1 = R_0;$$

$$D_2 = D_{gac} = -\frac{1}{\tau_{gac}^3} + \frac{R_2}{\tau_{gac}^2} - \frac{R_1}{\tau_{gac}} + R_0;$$

$$D_3 = D_{gde} = -\frac{1}{\tau_{gde}^3} + \frac{R_2}{\tau_{gde}^2} - \frac{R_1}{\tau_{gde}} + R_0;$$

$$D_4 = D_{gn1} = \frac{1}{\tau_{gn1}^3} + \frac{R_2}{\tau_{gn1}^2} + \frac{R_1}{\tau_{gn1}} + R_0;$$

$$D_5 = D_{gn2} = -\frac{1}{\tau_{gn2}^3} + \frac{R_2}{\tau_{gn2}^2} - \frac{R_1}{\tau_{gn2}} + R_0;$$

$$D_6 = D_{gn3} = -\frac{1}{\tau_{gn3}^3} + \frac{R_2}{\tau_{gn3}^2} - \frac{R_1}{\tau_{gn3}} + R_0;$$

$$D_7 = D_{tac} = -\frac{1}{\tau_{tac}^3} + \frac{R_2}{\tau_{tac}^2} - \frac{R_1}{\tau_{tac}} + R_0;$$

$$D_8 = D_{tde} = -\frac{1}{\tau_{tde}^3} + \frac{R_2}{\tau_{tde}^2} - \frac{R_1}{\tau_{tde}} + R_0;$$

$$D_9 = D_{tn1} = -\frac{1}{\tau_{tn1}^3} + \frac{R_2}{\tau_{tn1}^2} - \frac{R_1}{\tau_{tn1}} + R_0;$$

$$D_{10} = D_{tn2} = \frac{1}{\tau_{tn2}^3} + \frac{R_2}{\tau_{tn2}^2} + \frac{R_1}{\tau_{tn2}} + R_0;$$

$$D_{11} = D_{tn3} = -\frac{1}{\tau_{tn3}^3} + \frac{R_2}{\tau_{tn3}^2} - \frac{R_1}{\tau_{tn3}} + R_0.$$

Note that the signs for D_{gn1} and D_{tn2} are different from the others terms due to their positive expressions in the exponential term.

Natural Response The first step in solving for the natural response, $\theta_n(t)$, is to determine the roots of the characteristic equation from Eq. (1.26), $s^3 + P_2s^2 + P_1s + P_0 = 0$. Using typical parameter values described later, the system has one real root α_1 and one pair of complex roots, $\alpha_2 \pm j\beta_2$. Thus, using superposition with the delayed inputs, $\theta_n(t)$ is

$$
\begin{aligned}
\theta_n(t) = &\left(K_{11}e^{\alpha_1 t} + K_{12}e^{\alpha_2 t}\cos(\beta_3 t) + K_{13}e^{\alpha_2 t}\sin(\beta_2 t)\right)u(t) \\
&+\begin{pmatrix} K_{21}e^{\alpha_1(t-5\tau_{gn1})} + K_{22}e^{\alpha_2(t-5\tau_{gn1})}\cos\left(\beta_3\left(t-5\tau_{gn1}\right)\right) \\ +K_{23}e^{\alpha_2(t-5\tau_{gn1})}\sin\left(\beta_2\left(t-5\tau_{gn1}\right)\right) \end{pmatrix} u\left(t-5\tau_{gn1}\right) \\
&+\begin{pmatrix} K_{31}e^{\alpha_1(t-5T_1)} + K_{32}e^{\alpha_2(t-5T_1)}\cos\left(\beta_3\left(t-5T_1\right)\right) \\ +K_{33}e^{\alpha_2(t-5T_1)}\sin\left(\beta_2\left(t-5T_1\right)\right) \end{pmatrix} u\left(t-5T_1\right) \\
&+\begin{pmatrix} K_{41}e^{\alpha_1(t-5T_2)} + K_{42}e^{\alpha_2(t-5T_2)}\cos\left(\beta_3\left(t-5T_2\right)\right) \\ +K_{43}e^{\alpha_2(t-5T_2)}\sin\left(\beta_2\left(t-5T_2\right)\right) \end{pmatrix} u\left(t-5T_2\right) \\
&+\begin{pmatrix} K_{51}e^{\alpha_1(t-5T_3)} + K_{52}e^{\alpha_2(t-5T_3)}\cos\left(\beta_3\left(t-5T_3\right)\right) \\ +K_{53}e^{\alpha_2(t-5T_3)}\sin\left(\beta_2\left(t-5T_3\right)\right) \end{pmatrix} u\left(t-5T_3\right) \\
&+\begin{pmatrix} K_{61}e^{\alpha_1(t-T_3-5\tau_{tn2})} + K_{62}e^{\alpha_2(t-T_3-5\tau_{tn2})}\cos\left(\beta_3\left(t-T_3-5\tau_{tn2}\right)\right) \\ +K_{63}e^{\alpha_2(t-T_3-5\tau_{tn2})}\sin\left(\beta_2\left(t-T_3-5\tau_{tn2}\right)\right) \end{pmatrix} u\left(t-T_3-5\tau_{tn2}\right) \\
&+\left(K_{71}e^{\alpha_1(t-T_4)} + K_{72}e^{\alpha_2(t-T_4)}\cos\left(\beta_3\left(t-T_4\right)\right)+K_{73}e^{\alpha_2(t-T_4)}\sin\left(\beta_2\left(t-T_4\right)\right)\right)u\left(t-T_4\right)
\end{aligned}
\tag{1.43}
$$

where K_{ij} are the constants to be determined from the initial conditions.

Complete Response   The complete response, $\theta(t)$, is found by summing Eqs. (1.42) and (1.43), $\theta(t) = \theta_n(t) + \theta_f(t)$, giving

$$
\begin{aligned}
\theta(t) = &\left(K_{11}e^{\alpha_1 t} + K_{12}e^{\alpha_2 t}\cos(\beta_3 t) + K_{13}e^{\alpha_2 t}\sin(\beta_2 t)\right)u(t) \\
&+\begin{pmatrix} K_{21}e^{\alpha_1(t-5\tau_{gn1})} + K_{22}e^{\alpha_2(t-5\tau_{gn1})}\cos\left(\beta_3\left(t-5\tau_{gn1}\right)\right) \\ +K_{23}e^{\alpha_2(t-5\tau_{gn1})}\sin\left(\beta_2\left(t-5\tau_{gn1}\right)\right) \end{pmatrix} u\left(t-5\tau_{gn1}\right) \\
&\qquad\qquad\vdots \\
&+\begin{pmatrix} K_{71}e^{\alpha_1(t-T_4)} + K_{72}e^{\alpha_2(t-T_4)}\cos\left(\beta_3\left(t-T_4\right)\right) \\ +K_{73}e^{\alpha_2(t-T_4)}\sin\left(\beta_2\left(t-T_4\right)\right) \end{pmatrix} u\left(t-T_4\right) \\
&+\delta\left(A_{1,1} + A_{1,2}e^{-t/\tau_{gac}} + A_{1,3}e^{-t/\tau_{gde}} + \cdots + A_{1,11}e^{-t/\tau_{tn3}}\right)u(t) \\
&+\delta\begin{pmatrix} A_{2,1} + A_{2,2}e^{-(t-5\tau_{gn1})/\tau_{gac}} + A_{2,3}e^{-(t-5\tau_{gn1})/\tau_{gde}} \\ +\cdots + A_{2,11}e^{-(t-5\tau_{gn1})/\tau_{tn3}} \end{pmatrix} u\left(t-5\tau_{gn1}\right) \\
&\qquad\qquad\vdots \\
&+\delta\begin{pmatrix} A_{7,1} + A_{7,2}e^{-(t-T_4)/\tau_{gac}} + A_{7,3}e^{-(t-T_4)/\tau_{gde}} \\ +\cdots + A_{7,11}e^{-(t-T_4)/\tau_{tn3}} \end{pmatrix} u\left(t-T_4\right)
\end{aligned}
\tag{1.44}
$$

$$
= \left(\begin{array}{c} K_{11}e^{\alpha_1 t} + K_{12}e^{\alpha_2 t}\cos(\beta_3 t) + K_{13}e^{\alpha_2 t}\sin(\beta_2 t) \\ +\delta\left(A_{1,1} + A_{1,2}e^{-t/\tau_{gac}} + A_{1,3}e^{-t/\tau_{gde}} + \cdots + A_{1,11}e^{-t/\tau_{tn3}} \right) \end{array} \right) u(t)
$$

$$
+ \left(\begin{array}{c} K_{21}e^{\alpha_1 (t-5\tau_{gn1})} + K_{22}e^{\alpha_2 (t-5\tau_{gn1})}\cos\left(\beta_3\left(t - 5\tau_{gn1}\right)\right) \\ +K_{23}e^{\alpha_2 (t-5\tau_{gn1})}\sin\left(\beta_2\left(t - 5\tau_{gn1}\right)\right) \\ +\delta\left(\begin{array}{c} A_{2,1} + A_{2,2}e^{-(t-5\tau_{gn1})/\tau_{gac}} + A_{2,3}e^{-(t-5\tau_{gn1})/\tau_{gde}} \\ + \cdots + A_{2,11}e^{-(t-5\tau_{gn1})/\tau_{tn3}} \end{array} \right) \end{array} \right) u\left(t - 5\tau_{gn1}\right)
$$

$$
\vdots
$$

$$
+ \left(\begin{array}{c} K_{71}e^{\alpha_1 (t-T_4)} + K_{72}e^{\alpha_2 (t-T_4)}\cos\left(\beta_3\left(t - T_4\right)\right) \\ +K_{73}e^{\alpha_2 (t-T_4)}\sin\left(\beta_2\left(t - T_4\right)\right) \\ +\delta\left(\begin{array}{c} A_{7,1} + A_{7,2}e^{-(t-T_4)/\tau_{gac}} + A_{7,3}e^{-(t-T_4)/\tau_{gde}} \\ + \cdots + A_{7,11}e^{-(t-T_4)/\tau_{tn3}} \end{array} \right) \end{array} \right) u\left(t - T_4\right).
$$

Next, we determine the unknown constants K_{11} to K_{73}, twenty-one terms in total. At the beginning of a saccade, the system is in steady-state and all initial conditions are assumed to be zero. Using the initial conditions, that is, $\theta(0) = 0$ (initial displacement), $\dot{\theta}(0) = 0$ (initial velocity) and $\ddot{\theta}(0) = 0$ (initial acceleration), and Eq. (1.44), we generate three equations to determine three of the unknown constants, K_{11}, K_{12}, and K_{13}, from

$$
\begin{aligned}
\theta(0) &= 0 = K_{11} + K_{12} + A_{1,1} + A_{1,2} + \cdots + A_{1,11} \\
\dot{\theta}(0) &= 0 = \alpha_1 K_{11} + \alpha_2 K_{12} + \beta_2 K_{13} - \left(\frac{A_{1,2}}{\tau_{gac}} + \cdots + \frac{A_{1,11}}{\tau_{tn3}} \right) \\
\ddot{\theta}(0) &= 0 = \alpha_1 K_{11} + \left(\alpha_2^2 - \beta a_2^2 \right) K_{12} + 2\alpha_2 \beta_2 K_{13} + \frac{A_{1,2}}{\tau_{gac}^2} + \cdots + \frac{A_{1,11}}{\tau_{tn3}^2}.
\end{aligned} \tag{1.45}
$$

The other unknown constants, K_{21} to K_{73}, are determined at each of the switch times, and the fact that the variables $\theta(t)$, $\dot{\theta}(t)$ and $\ddot{\theta}(t)$ must be continuous in time. For instance, at switch time T_1, $\theta\left(T_1^-\right) = \theta\left(T_1^+\right)$, $\dot{\theta}\left(T_1^-\right) = \dot{\theta}\left(T_1^+\right)$ and $\ddot{\theta}\left(T_1^-\right) = \ddot{\theta}\left(T_1^+\right)$. Thus, at time T_1, we have

$$
\begin{aligned}
\theta(T_1) &= K_{11}e^{\alpha_1 T_1} + \cdots + A_{2,11}e^{-(T_1-5\tau_{gn1})/\tau_{tn3}} \\
&= K_{11}e^{\alpha_1 T_1} + \cdots + A_{2,11}e^{-(T_1-5\tau_{gn1})/\tau_{tn3}} + K_{31} + K_{32} + A_{3,1} + A_{3,2} + \cdots + A_{3,11}
\end{aligned}
$$

$$
\begin{aligned}
\dot{\theta}(T_1) &= \alpha_1 K_{11}e^{\alpha_1 T_1} + \cdots - \frac{A_{2,11}e^{-(T_1-5\tau_{gn1})/\tau_{tn3}}}{\tau_{tn3}} \\
&= \alpha_1 K_{11}e^{a_1 T_1} + \cdots - \frac{A_{2,11}e^{-(T_1-5\tau_{gn1})/\tau_{tn3}}}{\tau_{tn3}} + \alpha_1 K_{31} + \alpha_2 K_{32} \\
&\quad + \beta_2 K_{33} - \left(\frac{A_{3,2}}{\tau_{gac}} + \cdots + \frac{A_{3,11}}{\tau_{tn3}} \right)
\end{aligned} \tag{1.46}
$$

$$\dddot{\theta}(T_1) = \alpha_1^2 K_{11} e^{\alpha_1 T_1} + \cdots + \frac{A_{2,11} e^{-(T_1 - 5\tau_{gn1})/\tau_{tn3}}}{\tau_{tn3}^2}$$

$$= \alpha_1^2 K_{11} e^{\alpha_1 T_1} + \cdots + \frac{A_{2,11} e^{-(T_1 - 5\tau_{gn1})/\tau_{tn3}}}{\tau_{tn3}^2} + \alpha_1 K_{31} + \left(\alpha_2^2 - \alpha_2^2\right) K_{32}$$

$$+ 2\alpha_2 \beta_2 K_{33} + \frac{A_{32}}{\tau_{gac}^2} + \cdots + \frac{A_{3,11}}{\tau_{tn3}^2}$$

which requires

$$0 = K_{31} + K_{32} + A_{3,1} + A_{3,2} + \cdots + A_{3,11}$$

$$0 = \alpha_1 K_{31} + \alpha_2 K_{32} + \beta_2 K_{33} - \left(\frac{A_{3,2}}{\tau_{gac}} + \cdots + \frac{A_{3,11}}{\tau_{tn3}}\right) \tag{1.47}$$

$$0 = \alpha_1 K_{31} + \left(\alpha_2^2 - \beta_2^2\right) K_{32} + 2\alpha_2 \beta_2 K_{33} + \frac{A_{32}}{\tau_{gac}^2} + \cdots + \frac{A_{3,11}}{\tau_{tn3}^2}$$

The constants K_{31}, K_{32}, and K_{33} are solved from Eq. (1.47). The equations for the other unknown constants at the switch times follow similarly.

Example 1.1
Consider the oculomotor system shown in Fig. 1.1. Given the initial conditions and parameter below, create a Simulink program and plot the neural inputs, actives state tensions, position, velocity and acceleration.

$\theta(0) = 0°$, $\dot{\theta}(0) = 0°\text{s}^{-1}$, $\ddot{\theta}(0) = \ddot{0}°\text{s}^{-2}$, $T_1 = 0.0044$ s, $T_2 = 0.0259$ s, $T_3 = 0.0293$ s, $T_4 = 0.0462$ s, $F_{p1} = 1.06$ N, $F_{p2} = 0.9331$ N, $F_{p3} = 0.3790$ N, $F_{g0} = 0.4$ N, $F_{gs} = 0.5546$ N, $F_{t0} = 0.4$ N, $F_{ts} = 0.2895$ N, $\tau_{gn1} = 0.000287$ s, $\tau_{gn2} = 0.0034$ s, $\tau_{gn3} = 0.0042$ s, $\tau_{gac} = 0.0112$ s, $\tau_{tn1} = 0.000939$ s, $\tau_{tn2} = 0.0012$ s, $\tau_{tn3} = 0.001$ s, $\tau_{tac} = 0.0093$ s, $\tau_{tde} = 0.0048$ s, $K_{se} = 124.9582$ Nm, $K_{lt} = 60.6874$ Nm, $K = 16.3597$ Nm, $B_1 = 5.7223$ Nms^{-1}, $B_2 = 0.5016$ Nms^{-1}, $B = 0.327$ Nms^{-1}, $J = 0.0022$ Nms^{-1}, and radius $= 0.0118$ m.

Solution
We first compute the following intermediate results

```
b12=b1+b2=6.2239
kst=kse+klt=185.6456
c3=b12*j=0.0137
c2=b12*bp+kst*j+2*b1*b2=8.1855
c1=b12*kp+kst*bp+2*(b2*klt+b1*kse)= 1.6535e+003
```

```
c0=kst*kp+2*kse*klt=1.8204e+004
delta=57.296/(r*c3)= 3.5397e+005
p2=c2/c3=596.7159
p1=c1/c3=1.2054e+005
p0=c0/c3= 1.3271e+006
```

The Simulink model is shown in Fig. 1.6. The diagram in (A) is implemented using Eq. (1.26). The input to the oculomotor plant is shown in diagram (B), with the agonist and antagonist active state tensions shown in (C) and (D) based on Eqs. (1.29) and (1.30). The neural input is based on Eqs. (1.27) and (1.28).

In Fig. 1.7 are plots of position, velocity, acceleration, agonist neural input and active state tension, and antagonist neural input and active state tension. From the antagonist neural input and active state tension plot, it is clear that the eye movement has post-saccade behavior. Peak return velocity is $-20°s^{-1}$ in Fig. 1.7 (B), which makes it a glissade.

1.3   PARAMETER ESTIMATION AND SYSTEM IDENTIFICATION

The model presented here involves a total of 25 parameters describing the oculomotor plant, neural inputs and active-state tensions that are estimated by system identification as described in the next section. Initial estimates of the model parameters are important since they affect the convergence of the estimation routine. In this model, the initial estimates are derived from previously published experimental observations and briefly reviewed here. A more detailed discussion of the parameter estimates for human and monkey are given in Sections 1.4 and 1.5.

The oculomotor plant parameters for a human are determined based on the work by Enderle, J. (2002, 1988); Zhou et al. (2009); Enderle et al (1991). The eyeball moment of inertia, J, has an initial estimated value of $2.2 \times 10^{-3} Ns^2/m$, assuming the radius of the eyeball is 11 mm. The parameter values of K_{se}, K_{lt}, B_1, B_2 are based on a linear muscle model (Enderle et al, 1991). From the length-tension curves, $K_{se} = 125$ N/m, and $K_{lt} = 60.7$ N/m. B_1 is selected as 5.6 Ns/m, and B_2 is 0.5 Ns/m to fit the nonlinear force-velocity relationship. K is determined by steady-state analysis of the model, yielding a value of 16.34 N/m. B is determined by considering the dominant orbital time constant 0.02 s, which yields $B = 0.327$ Ns/m.

The initial estimated start time, T_p, of the saccade is based on selecting a threshold from the estimated velocity from the position data. Such estimations can also be obtained using a first-spike method as reported by others.

The nine parameters for the agonist neural input and active-state tension are T_1, T_2, F_{p1}, F_{p2}, with time constants $\tau_{gn1}, \tau_{gn2}, \tau_{gn3}, \tau_{gac}, \tau_{gde}$. The initial estimate of the duration, T_1, is assumed to be 3 ms. $F_{p2}, \tau_{gn2},$ and τ_{gn3} are empirically estimated from EBN firing rates of the monkey.

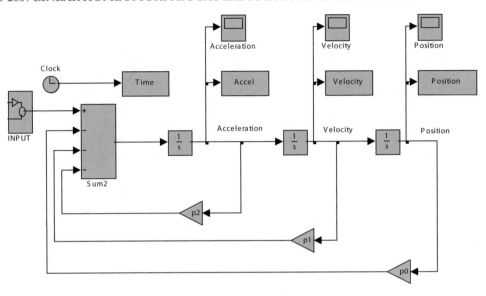

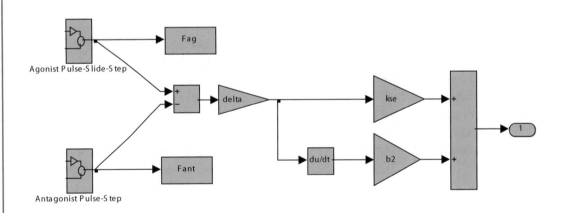

Figure 1.6: Simulink program for Example 1.1. (A) Main program. (B) Input to plant.

F_{p1}, τ_{gac}, and $T_2 - T_1$ are estimated from the peak velocity of the data, using the method reported by Enderle and Wolfe (1988).

The eight parameters for antagonist neural input and active-state tension are T_3, T_4, F_{p3}, with time constants τ_{tn1}, τ_{tn2}, τ_{tn3}, τ_{tac}, τ_{tde}. The initial estimates for time constants τ_{tn1}, τ_{tn2}, and τ_{tn3} are 2 ms, 1.5 ms, and 0.2 ms, respectively. The antagonist onset delay, $T_3 - T_2$, is the time between the antagonist step and agonist pulse, and it is variable from saccade to saccade. Physiological

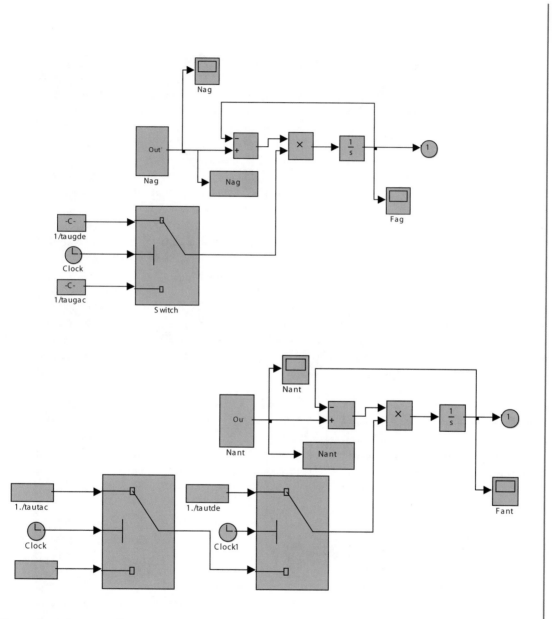

Figure 1.6: (continued). Simulink program for Example 1.1. (C) Agonist active state tension. (D) Antagonist active state tension.

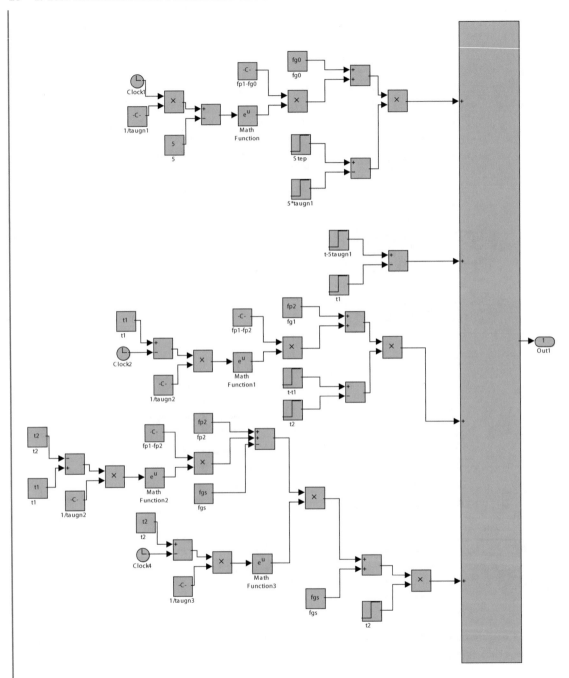

Figure 1.6: (continued). Simulink program for Example 1.1. (E) Agonist neural input.

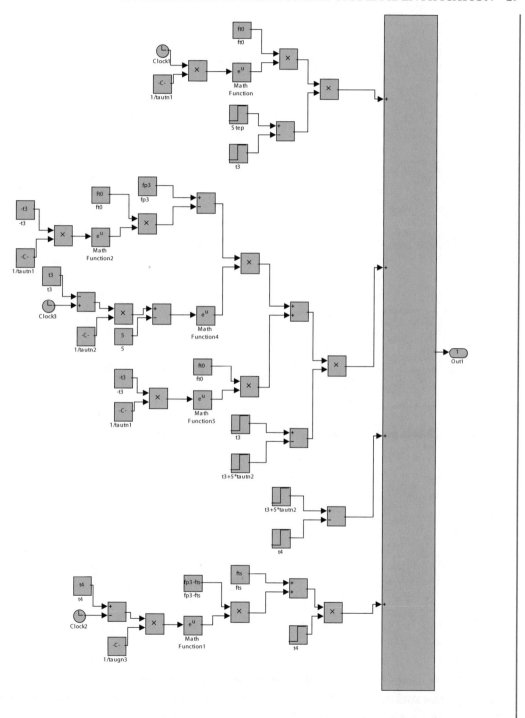

Figure 1.6: (continued). Simulink program for Example 1.1. (F) Antagonist neural input.

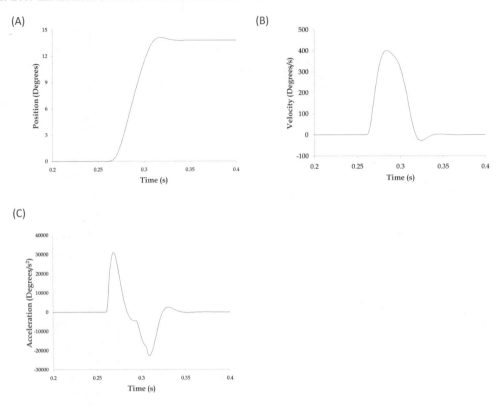

Figure 1.7: Plots of position, velocity and acceleration for Example 1.1.

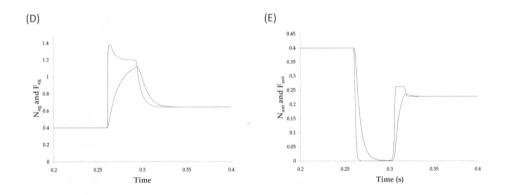

Figure 1.7: (continued). Plots of agonist neural input and active state tension, and antagonist neural input and active state tension for Example 1.1, (D) and (E). Active state tension is drawn with a blue line and the neural input is drawn with a red line.

observations suggest that the delay varies from 3 ms to 20 ms for large saccades (Robinson, D., 1981). Without losing generality, the antagonist onset time, T_3, is estimated as a time between T_2 and a time near the end of the saccade.

A second peak velocity is observed in saccades with either a dynamic or glissadic overshoot. Data suggest that dynamic overshoot has a higher second peak velocity than glissadic overshoot. In fact, we show that the prominent second peak velocity may be caused by the antagonist PIRB. The antagonist parameters F_{p3}, T_4, and τ_{tac} control the rebound burst and are determined by analyzing the relationships between them and the second peak velocity.

Steady-state tensions F_{g0}, F_{gs}, F_{t0} and F_{ts}, are determined as functions of eye position at steady-state

$$F = \begin{cases} 0.4 + 0.0175\,|\theta| & \text{N} \quad \text{for} \quad \theta \geq 0° \\ 0.4 - 0.0125\,|\theta| & \text{N} \quad \text{for} \quad \theta < 0° \end{cases}$$

where the eye position θ is always positive for the agonist and negative for the antagonist active-state tension. These four parameters are fixed for a particular saccade during the estimation routine.

1.3.1 SYSTEM IDENTIFICATION

In a time domain system identification method, a conjugate gradient descent algorithm is applied directly to the objective function. The system identification problem is stated as follows:

Find a parametric vector $\mathbf{y} = (y_1, y_2, \ldots, y_n)$ of the oculomotor plant to minimize (optimize) an objective function $f = f(\mathbf{y})$ subject to equality constraints $h_i(\mathbf{y}) = 0, i = 1 \ldots q$ and inequality constraints $g_i(\mathbf{y}) \leq 0, i = 1 \ldots m$.

Here the parametric vector $\mathbf{y}$ is composed of the 25 parameters that need to be estimated. Let $\theta_M(\mathbf{y}, t)$ be the solution of the homeomorphic oculomotor plant as defined by Eq. (1.26). The analytical form of $\theta_M(\mathbf{y}, t)$ is derived by substituting the analytical solutions of active-state tensions and using the separation of homogeneous and particular solutions.

Next, let $p_1, \ldots p_N$ be a set of experimental data points of eye position in time domain $p_i = (t_i, \theta_i)$, where t_i is the ith time point and θ_i is the ith eye position in degrees. The 2009 linear homeomorphic model is defined by its own parametric vector $\mathbf{a} = (P_2, P_1, P_0, \delta, K_{se}, B_2, F_{ag}, F_{ant})$

$$f(P, \mathbf{a}) = \dddot{\theta} + P_2\ddot{\theta} + P_1\dot{\theta} + P_0\theta - \left(\delta(K_{se}\left(F_{ag} - F_{ant}\right) + B_2\left(\dot{F}_{ag} - \dot{F}_{ant}\right)\right) = 0 \quad (1.48)$$

where $P = \bigcup_{i \in H} p_i$ which includes all possible time-position points during saccade.

The goal is to find $\mathbf{x}^*$ that satisfies

$$\mathbf{x}^* = \min \arg \sum_{i=1}^{N} [D(p_i, \mathbf{a})]^2 \quad (1.49)$$

where $D(p_i, \mathbf{a})$ is a suitable distance[3]. Here the distance is defined as the algebraic distance of experimental data point p_i from the model $|f(p_i, \mathbf{a})|$. Thus, Eq. (1.49) becomes

$$\mathbf{x}^* = \min \arg \sum_{i=1}^{N} |f(p_i, \mathbf{a})|^2 . \tag{1.50}$$

The vector $\mathbf{a}$ in Eq. (1.50) is then represented by the parameter vector $\mathbf{x}$, and thus the objective function is the following:

$$F(\mathbf{x}) = \sum_{i=1}^{N} |f(p_i, \mathbf{x})|^2 \tag{1.51}$$

subject to:

1. *Equality constraints.* The initial and final positions should satisfy

$$h_1 = K_{se}(F_{ag}(t = 0) - F_{ant}(t = 0)) - P_0\theta(t = 0) = 0,$$
$$h_2 = K_{se}(F_{ag}(t = t_s) - F_{ant}(t = t_s)) - P_0\theta(t = t_s) = 0$$

   where at time t_s the saccade ends.

2. *Inequality constraints.* For parameters $(K_{se}, K_{lt}, B_1, B_2, K_p, B_p, J)$, inequality constraints are added to limit variations within a desirable range around their initial estimations. Those estimations are obtained from experimental data. The purpose of the inequality constraints is to eliminate the possibility of abnormal values from the gradient method. Here, we set a constraint within 25% around the initial guess.

$$\begin{cases} g_i = x_i/x_i^0 - (1 + 25\%) \le 0 \\ g_{i+1} = (1 - 25\%) - x_i/x_i^0 \le 0 \end{cases}$$

for each parameter listed above. This constrained optimization problem is transformed into an unconstrained problem using a transformation function of the form:

$$\varphi(\mathbf{x}, \mathbf{r}) = F(\mathbf{x}) + P(\mathbf{h}(\mathbf{x}), \mathbf{g}(\mathbf{x}), \mathbf{r}) \tag{1.52}$$

where $\mathbf{r}$ is a vector of penalty parameters and P is a real valued function whose action is imposing the penalty on the objective function controlled by $\mathbf{r}$. P is defined by the following quadratic loss function:

$$P(\mathbf{h}(\mathbf{x}), \mathbf{g}(\mathbf{x}), \mathbf{r}) = r_1 \sum_{i=1}^{p} [h_i(\mathbf{x})]^2 + r_2 \sum_{i=1}^{m} \log(-g_i(\mathbf{x})) . \tag{1.53}$$

Here the log barrier function methods are applied to the inequality constraints. In fact, the function becomes infinite if any of the inequalities are active. When the iterative process is started from a feasible point (initial guess), it cannot go into the infeasible region because the iterative process cannot cross the barrier. If $h_i(\mathbf{x}) \ne 0$, Eq. (1.53) also gives a positive value to the function P, and the cost function is penalized.

---

[3]In the frequency method, this distance is defined as error squared between transfer function $G(jw, b)$ and the plant frequency response $A(w)e^{j\theta(w)}$ as carried out in Enderle, J., 1988.

1.3.2   NUMERICAL GRADIENT

One shortcoming of time domain method is that the exact gradient cannot be evaluated as in the frequency response method (Enderle and Wolfe, 1988). This is because the parameter vector in the objective function, Eq. (1.49), is implicit and the derivatives are impossible to calculate directly.

At a first glance, a simple one-step 1st or 2nd order finite differencing method appears plausible. However, due to truncation and round off error in Taylor series expansion and the parameters with different scales, this method makes the parameter values unstable and without convergence. Here, the finite differencing is calculated with smaller and smaller finite values of the step h, to $h \to 0$. By the use of Neville's algorithm, each finite differencing calculation produces both an extrapolation of higher order, and it also extrapolations of previous lower orders but with smaller scales h.

The algorithm is based on Ridders, with some necessary changes. The derivatives are evaluated in the objective function with respect to each parameter individually. The initial step length is estimated by

$$h = eps * \max(|parm|, 1/s) * sign(parm) \tag{1.54}$$

where s is a large number (i.e., 1000) and eps is a small number.

1.3.3   VELOCITY AND ACCELERATION ESTIMATION

Velocity and acceleration are essential in the study of the oculomotor system. They are computed from data by the central difference method. To reduce aliasing, a suitable spread for central differencing must be estimated. Based on the relationship between the bandwidth and the amount of spread between points

$$\text{bandwidth (Hz)} = \frac{0.443 \times \text{sampling rate (Hz)}}{\text{spread}}.$$

The maximum frequency for saccade velocity is estimated at 74 Hz, and 45 Hz for saccade acceleration (Bahill, A., 1980). Thus, for the saccade which has a sampling rate of 1000 Hz, the spread for saccade velocity is 6, and the spread for saccade acceleration is estimated as 8. The velocity is calculated as

$$\dot{y}(kT) = \frac{y((k+3)T) - y((k-3)T)}{6T}$$

which has a bandwidth of 74 Hz. The acceleration is estimated as

$$\ddot{y}(kT) = \frac{\dot{y}((k+4)T) - \dot{y}((k-4)T)}{8T}$$

which has a bandwidth of 55 Hz.

For the data sampled at 2000 Hz, the spread for saccade velocity is 12, and the spread for saccade acceleration is estimated as 16. The velocity is estimated as,

$$\dot{y}(kT) = \frac{y((k+6)T) - y((k-6)T)}{12T}$$

which has a bandwidth of 74 Hz. And the acceleration is estimated as the following:

$$\ddot{y}(kT) = \frac{\dot{y}((k+8)T) - \dot{y}((k-8)T)}{16T}$$

which has a bandwidth of 55 Hz.

1.3.4   INVERSE FILTER

The inverse filter is a discrete 1st order low-pass filter (LPF) that filters the neuron bursting rate to active-state tension. Consider the EBN firing rate, where in the continuous time domain, the LPF has the form of

$$\dot{F}_{ag} = \frac{N_{ag} - F_{ag}}{\tau_{ag}}, \text{ or } \tau_{ag}\dot{F}_{ag} + F_{ag} = N_{ag} . \tag{1.55}$$

Since the EBN firing rate is discontinuous, the differential equation can be discretised using the approximation:

$$\frac{dF_{ag}(t)}{dx} \approx \frac{F_{ag,k} - F_{ag,k-1}}{T_s} \tag{1.56}$$

where T_s is the interval between each measurement, i.e., the sampling interval 0.0005s at frequency of 2000 Hz. Thus, the differential equation representing the 1st order low pass filter is converted to

$$\tau_{ag}\frac{F_{ag,k} - F_{ag,k-1}}{T_s} + F_{ag,k} = N_{ag,k} . \tag{1.57}$$

Simplification and re-arrangement of Eq. (1.56) gives

$$F_{ag,k} = \left(\frac{\tau_{ag}}{\tau_{ag} + T_s}\right)F_{ag,k-1} + \left(\frac{T_s}{\tau_{ag} + T_s}\right)N_{ag,k} . \tag{1.58}$$

Here $F_{ag,k}$ and $F_{ag,k-1}$ are the agonist active-state tensions at time point k and $k - 1$, respectively. $N_{ag,k}$ is the burst frequency at sampling time point k, T_s is the sampling interval, and $\tau_{ag} = \tau_{gac}(u(t) - u(t - T_2)) + \tau_{gde}u(t - T_2)$.

We transform the neuron firing rate in Hz to active-state tension in N with the following normalization

$$\frac{H - H_0}{F - F_{g0}} = S \left(\frac{Hz}{N}\right) \tag{1.59}$$

where H is the firing rate, H_0 is steady-state firing rate before the saccade, F is the active state tension, F_{g0} is the initial active-state tension, and S is the coefficient in $\left(\frac{Hz}{N}\right)$. This equation scales the low-pass filtered neuron firing rate to an active state tension. A coefficient selected based on physiological evidence is

$$S = \frac{100}{0.4} \tag{1.60}$$

which is a reasonable approximation (Enderle, J., 2002). Practically, S and H_0 are quite variable for saccades of different sizes, and are therefore selected manually to match the data. This is because the firing data from a single neuron may not represent the average firing of all the neurons during a saccade. As a result, the observation from the firing of a single or several neurons is not accurate for general use in the generation of saccades.

1.4   INITIAL PARAMETER ESTIMATION FOR HUMANS

Suitable initial estimates for model parameters are important for system identification accuracy. To meet local convergence requirements of conjugate gradient algorithm, a reasonable initial guess is required for a success in iteration process. This section describes the algorithms used in calculating initial parameter estimates.

1.4.1   ESTIMATION OF THE START TIME AND DURATION OF A SACCADE

As a first step, the start time and duration of a saccade are estimated by using thresholds. The velocity profile of a saccade with glissadic overshoot is shown in Fig. 1.8. To avoid the detection error due to noise before filtering, *Threshold 1* is set at a relatively large value (i.e., $70°\text{s}^{-1}$). The estimate of

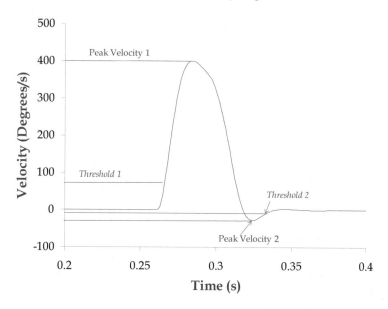

Figure 1.8: Velocity profile for a typical saccade with glissadic overshoot.

the saccade start time is 6 ms before the time when the velocity reaches *Threshold 1*. *Threshold 2* represents the end of the saccade and is set at $7°\text{s}^{-1}$. The saccade end time is 6 ms after the time when the velocity falls below *Threshold 2*.

Obviously, these two thresholds are artificially specified and do not have precise relationships with the real start time and end time of the saccade. These estimates are required since the Kaiser window filter used in the estimation routine needs them. After filtering, these two parameters are evaluated again. The saccade start time is chosen as a time when the velocity is nearing $0°\text{s}^{-1}$ and is about to increase. The saccade end time is when the agonist and antagonist active-state tensions reach steady-state at their final values.

1.4.2   ESTIMATION OF MODEL PARAMETERS

The fixed parameters for the model are determined first and held constant for all saccades across all subjects. J has a value of $2.2 \times 10^{-3} \frac{Ns^2}{m}$ when the radius of eyeball is 11 mm. From the length-tension curves, $K_{se} = 125 \frac{\text{N}}{\text{m}}$ and $K_{lt} = 60.7 \frac{\text{N}}{\text{m}}$. K is determined by steady-state analysis of the model as described using Eqs. (1.61) and (1.62). At the beginning of saccade, Eq. (1.26) is

$$K_{se}(F_{g0} - F_{t0}) = P_0\theta\,(0)$$
$$= (K_{st}K + 2K_{lt}K_{se})\theta\,(0) \tag{1.61}$$

and at the end of saccade

$$K_{se}(F_{gs} - F_{ts}) = P_0\theta\,(+\infty)$$
$$= (K_{st}K + 2K_{lt}K_{se})\theta\,(+\infty). \tag{1.62}$$

K should satisfy both Eqs. (1.61) and (1.62), yielding a value of $16.34 \frac{\text{N}}{\text{m}}$. The initial estimate for B is determined by using the orbital time constant of 0.02 s, which is

$$\frac{B}{K} = 0.02s \tag{1.63}$$

Thus, $B = 0.327 \frac{\text{Ns}}{\text{m}}$.

B_1 is a dominant parameter that strongly influences the peak velocity of saccades. For the input, the dominant parameters that affect the peak velocity are F_{p1} and τ_{gac}. The influence of F_{p1} and τ_{gac} on the peak velocity is plotted in Fig. 1.9 (A) τ_{gac} and (B) F_{p1}. Fig. 1.9 (A) illustrates the influence of B_1 and τ_{gac} on peak velocity with F_{p1} held constant. B_1 has a negative influence on peak velocity, while τ_{gac} has a positive influence on peak velocity. Figure 1.9 (B) shows the influence of B_1 and F_{p1} on peak velocity when τ_{ac} is 0.012s. As before, B_1 has a negative influence on peak velocity, while F_{p1} has a positive influence on peak velocity. Since the typical peak velocities for saccades range from $200°\text{s}^{-1}$ to $800°\text{s}^{-1}$, B_1 is selected as 5.6 Ns/m and B_2 is selected as 0.5 Ns/m. Several force-velocity curves for different values of B_1 are shown in Fig. 1.10. They maintain non-linear shape for the range of values from 2 to 6 since B_2 determines this property.

1.4.3   ESTIMATION OF PARAMETERS FOR THE AGONIST MUSCLE

The parameters for agonist muscle are T_1, T_2, F_{p1}, F_{p2}, and time constants τ_{gn1}, τ_{gn2}, τ_{gn3}, τ_{gac}, and τ_{gde}. The EBN burst firing has a minimum duration T_1 for saccades for all sizes (Enderle, J.,

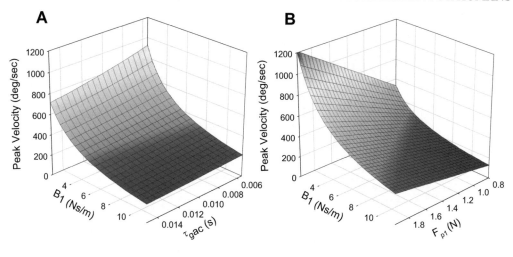

Figure 1.9: Influence of B_1, F_{p1} and τ_{gac} on the peak velocity of saccade. (A) F_{p1} is held constant at 1.2 N; (B) τ_{gac} is held constant at 0.012 s . (Note that F_{p2} is set equal to F_{p1}.)

2002). In the model, this duration is assumed to be 3 ms. F_{p2}, τ_{gn2} and τ_{gn3} are empirically estimated from observations of the EBN firing rate in monkey. F_{p2} is estimated as a function of F_{p1},

$$F_{p2} = F_{gs} + 0.75 \left(F_{p1} - F_{gs} \right) \tag{1.64}$$

τ_{gn2} and τ_{gn3} are estimated from the shape of firing rate and assumed to be dependent on the saccade magnitude. This dependency may be related to the different durations for saccades of different sizes.

From Fig. 1.11, the empirical equation for τ_{gn2} is

$$\tau_{gn2} = \min(0.00039 + 0.00034\,|\theta|\,,\,0.007) \tag{1.65}$$

and the equation for τ_{gn3} is

$$\tau_{gn3} = \min(0.0018 + 0.00026\,|\theta|\,,\,0.007) \tag{1.66}$$

F_{p1}, τ_{gac}, and $T_2 - T_1$ are estimated from peak velocity (the primary Peak Velocity 1 in Fig. 1.8). The time constants for antagonist τ_{tn1} and τ_{tde} also influence the peak velocity. We have assumed them to be 0.002 s and 0.005 s, respectively. Since T_3 is assumed to always be larger than T_2, the other antagonist parameters do not affect the peak velocity.

The time to peak velocity is estimated from

$$\left. \frac{\partial^2 \theta}{\partial t^2} \right|_{t=T_{mv}} = 0 \tag{1.67}$$

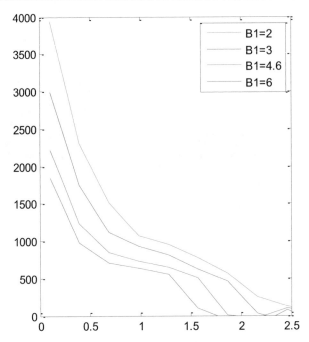

Figure 1.10: Force-velocity relationships for different values of B_1.

where at $t = T_{mv}$, the peak velocity is obtained. As an example, Fig. 1.12 illustrates the affect of agonist pulse magnitude, F_{p1}, and duration, $T_2 - T_1$, on Peak Velocity 1 under different activation time constants, τ_{ac}, for an adducting $8°$ saccade. In this figure, the range of F_{p1} comes from the fact that typical range of peak firing rate of motoneuron is 200-600 Hz. Accordingly, with 0.4 N $\sim$ 100 Hz, the range of F_p is typically from 0.8 N to 3 N.

T_{mv} for the saccade is around 16 ms. Increasing the agonist pulse magnitude increases Peak Velocity 1. Increasing pulse duration increases Peak Velocity 1 until $T_{mv} - T_1$, after which the duration does not have much influence on peak velocity. Note that due to the existence of minimum pulse duration T_1, the saccade could always achieve a velocity larger than $100°\text{s}^{-1}$. Increasing the agonist activation time constant, τ_{gac}, decreases the peak velocity as illustrated in Fig. 1.13. As expected, increasing the agonist activation time constant, τ_{gac}, decreases the peak velocity.

These observations provide three clues for estimating agonist pulse magnitude, F_{p1}, duration, $T_2 - T_1$, and the activation time constant, τ_{gac}:

- Pulse magnitude strongly influences peak velocity. Peak velocity increases while F_{p1} increases.

- When duration $T_2 - T_1$ is larger than threshold, it shows insignificant influences on peak velocity. The duration of agonist pulse is important for reaching the expected displacement of saccade.

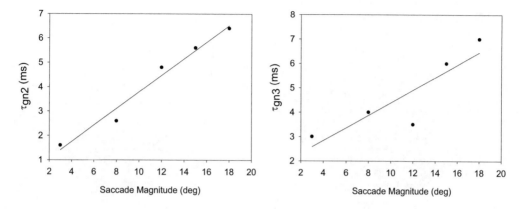

Figure 1.11: τ_{gn2} and τ_{gn3} are assumed to be dependent of the saccade magnitude. The lines are a linear regression result from the data points.

- Increasing the time constant τ_{gac} decreases the peak velocity.

According to the dynamic characteristics of the system and these observations, it is reasonable to roughly estimate T_2 equal to T_{mv} before final corrections. Due to the linear characteristic of the model, T_2 could be smaller or larger than T_{mv}. For a large saccade (i.e., the saccade magnitude is larger than 10°), T_2 is often larger than T_{mv} as observed in the data. T_2 can also be larger than T_{mv}, that necessary to reach the saccade destination without changing peak velocity.

Based on these observations, the agonist pulse magnitude F_{p1} and time constant τ_{gac} are determined by setting up a $F_{p1} - \tau_{gac}$ versus peak velocity look-up table which comes from the values of the nodes on the map shown in Fig. 1.13 and finding the most suitable Peak Velocity 1 in this table to fit the experimental data in the acceptable range of estimated peak acceleration. This matches the peak velocity of the model to that of the data. The time constant τ_{gn1} also has some influence on peak velocity; however, it primarily affects the peak acceleration.

As shown in Fig. 1.14, the agonist time constant, τ_{gn1}, is estimated from a look-up table to best match the peak acceleration. For the agonist, the decaying time constant, τ_{gde}, is assumed as 0.0065 s. Further, this parameter does not influence the time to the peak velocity or the peak velocity.

1.4.4   ESTIMATION OF PARAMETERS FOR ANTAGONIST MUSCLE

The parameters for the antagonist muscle are T_3, T_4, F_{p3}, and time constants τ_{tn1}, τ_{tn2}, τ_{tn3}, τ_{tac}, τ_{tde}. The time constants τ_{tn1}, τ_{tn2}, and τ_{tn3} are estimated with values of 2 ms, 1.5 ms, and 0.2 ms, respectively. The antagonist onset delay $T_3 - T_2$, is the delay between antagonist step (or rebound burst) and agonist pulse. This delay is variable for saccades. Physiological observations suggest that the delay varies from 3 ms to as long as 20 ms for large saccades (Robinson, D., 1981). Without losing generality, the antagonist onset time T_3, is estimated as a time between T_2 and the

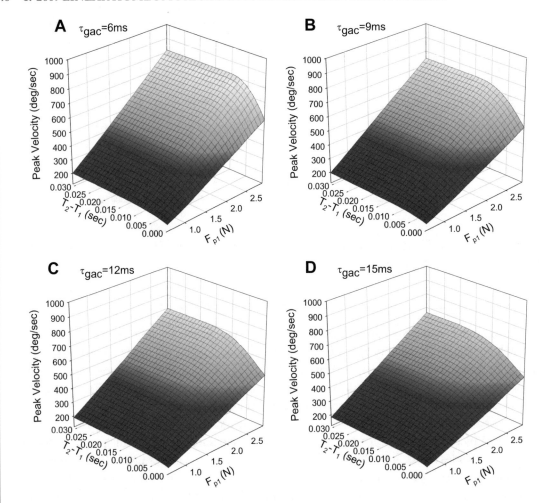

Figure 1.12: The affect of the agonist pulse magnitude, F_{p1} and duration, $T_2 - T_1$ on Peak Velocity 1 using different activation time constants, τ_{gac}.

time point near the end of saccade. This value is adjusted in the correction subroutine of initial estimation routine.

A second peak velocity (Peak Velocity 2 in Fig. 1.8) is observed in many saccades, those with dynamic and glissadic overshoot. Data suggest that saccades with dynamic overshoot often have higher Peak Velocity 2 than saccades with glissadic overshoot. In fact, the prominent second peak velocity mainly results from the antagonist rebound burst pulse. The agonist pulse duration has little influence on this peak velocity. The antagonist parameters F_{p3}, T_4, and τ_{tac} control the rebound burst and are determined by analyzing the relationships among them and Peak Velocity 2. Assuming that

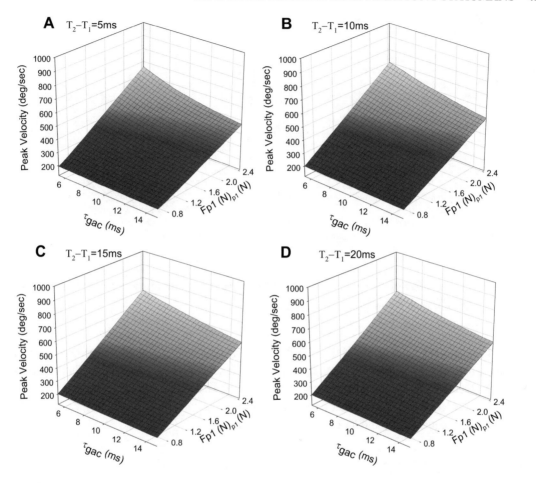

Figure 1.13: The affect of the agonist pulse magnitude F_{p1} and activation time constant τ_{gac} on Peak Velocity 1 using different durations $T_2 - T_1$.

τ_{tde} has little influence on the magnitude of Peak Velocity 2, Peak Velocity 2 is plotted in Fig. 1.15 under the influence of F_{p3}, T_4 and τ_{tac}.

Fig. 1.15 shows that τ_{tac} has a negative influence on the Peak Velocity 2. The rebound burst pulse magnitude F_{p3} strongly affects Peak Velocity 2. Peak Velocity 2 increases when the duration increases until saturation. As a matter of fact, since the second peak velocities are mostly under $-40°\text{s}^{-1}$ for saccades, the influence of τ_{tac} is limited to the range $\left(-40 \text{ to } 0°\text{s}^{-1}\right)$ and saturation is hardly achieved. In this range, a look-up table for F_{p3} and the rebound burst duration is used for Peak Velocity 2; one pair of F_{p3} and duration is selected from this table to match the Peak Velocity 2.

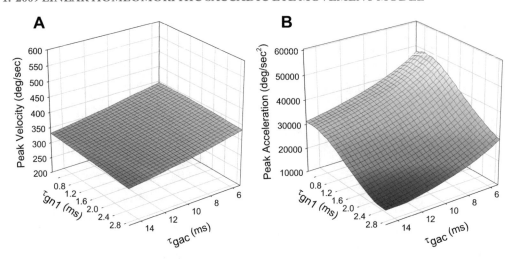

Figure 1.14: The affect of the agonist time constant τ_{gn1} and activation time constant τ_{gac} on (A) Peak Velocity 1 and (B) peak acceleration.

However, it should be noted that the Peak Velocity 2 is often small and is easily contaminated by noise. In some cases, a normal saccade can have a "fake" Peak Velocity 2, which is in fact due to noisy fluctuations. For such cases, the algorithm ignores Peak Velocity 2 since it is very small and does not calculate F_{p3} and the other parameters for the rebound burst. For normal saccades, F_{p3} is assumed to be equal to F_{ts} and T_4 is set to a large value.

1.4.5   CORRECTIONS

Some corrections are made at the end of initial estimation algorithm. The agonist pulse duration, T_2, and the antagonist onset delay, $T_3 - T_2$, are estimated again if the model generated saccade does not fit the target displacement. The algorithm increases T_2 and $T_3 - T_2$ into an acceptable range if the estimated final saccade displacement is smaller than the final displacement of data, and decreases T_2 and $T_3 - T_2$ if it is the contrary. Statistically, the correction of increasing T_2 and $T_3 - T_2$ mainly occurs in the case of large saccades due to their large displacement.

If T_2 and $T_3 - T_2$ are adjusted, the parameters for antagonist rebound burst are estimated again. The rebound burst duration is constrained between 3 ms and 14 ms according to physiological observations (Enderle, J., 2002). Finally, the start time of saccade is corrected to a small range around the previous estimate to achieve the smallest value of the objective function. This step is necessary because it is important to match the dynamics of the model to those of the data naturally or some abnormities occur.

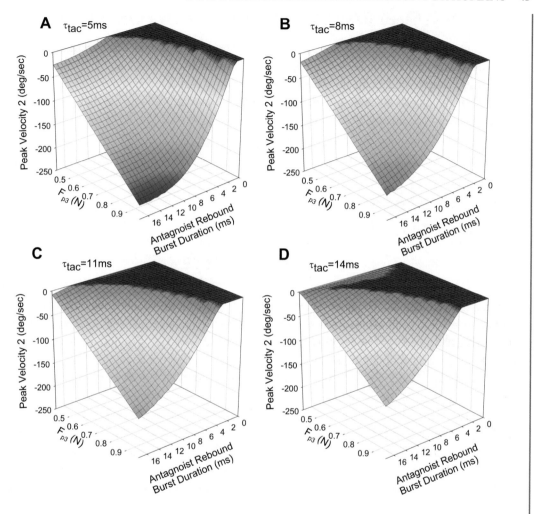

Figure 1.15: The affect of the antagonist rebound burst magnitude F_{p3} and activation time constant τ_{tac} on Peak Velocity 2 under different antagonist rebound burst durations $T_4 - T_3 - 5\tau_{tn2}$.

1.4.6 IMPLEMENTATION

In the implementation of algorithm, some parameters are transformed based on their physical meanings so that they always have a lower bound (i.e., the algorithm uses $F_{p1} - F_{gs}$ as a parameter instead of F_{p1} since F_{p1} is always larger than F_{gs}). The initial estimates are loaded into system identification subroutine and undergo an iteration process using the conjugate gradient method. For most cases studied, the initial guess algorithm demonstrates a good capability in providing a suitable initial estimate. In some cases, the initial guess is very close to the final result.

1.5 INITIAL PARAMETER ESTIMATION FOR MONKEY

Parameter estimates for monkey are based on data from the literature. The parameters of interest for monkey extraocular muscles is K_{se}, K_{lt}, B_1, B_2, the empirical equation of the static active-state tension and the parameters for the oculomotor plant J, B and K.

The radius of an average rhesus monkey (Macaca mulatta) eyeball is assumed as 10 mm (Fuchs and Luschei, 1971). Thus,

$$1g = 9.806 \times 10^{-3} \text{ N}$$
$$1° = 1.74 \times 10^{-4} \text{ m.} \tag{1.68}$$

1.5.1 STATIC CONDITIONS

In the static condition, muscle tension is given by

$$T = \frac{K_{se}}{K_{se} + K_{lt}} F - \frac{K_{se} K_{lt}}{K_{se} + K_{lt}} x_1 \tag{1.69}$$

(see Eq. (1.1) for details). From the length-tension data for rhesus monkey, we assume the length-tension data of rhesus monkeys are straight parallel lines in the operating region of muscle, with the slope estimated as 0.85 g/° = 47.90N/m. Assuming that $K_{se} = 125$N/m in the operating region of monkey extraocular muscles, with the slope of the length-tension curves 47.9N/m, and the slope given by in Eq. (1.69), $\frac{K_{se}K_{lt}}{K_{se}+K_{lt}}$, K_{lt} is estimated as 77.66 N/m. Fuchs et al. suggested that the extraocular muscle tension of monkey saturates with frequency above rates of $400° s^{-1}$, which is the upper curve in their Fig. 4 (Fuchs and Luschei, 1971). Thus, we assume this curve corresponds to the innervation level of $30° N$. At primary position, the tension is 44 g for $30° N$. The equilibrium point of the 2009 linear muscle model is estimated as 25° according to the passive elasticity, and the muscle length x_1 at primary position is 4.35 mm (human muscle is 3.705 mm). The empirical equation of the static active-state tension is estimated to match the monkey length-tension curves, and is

$$F = \begin{cases} 0.55 + 0.0175 |\theta|, & \text{for } \theta > 0 \\ 0.55 - 0.0125 |\theta|, & \text{for } \theta \leq 0. \end{cases}$$

The distribution of length-tension curves from the monkey data is unknown, thus, we assume that the distribution is similar to human as shown in Fig. 1.16. Comparing human static active-state tensions to monkey static active-state tensions, we see that the monkey's are generally larger, corresponding to higher firing frequency of motoneurons.

1.5.2 FORCE-VELOCITY CHARACTERISTICS

Parameters B_1 and B_2 describe the force-velocity characteristics of extraocular muscles. Widrick et al. (1997) compared the contractile properties of rat, rhesus monkey, and human type I muscle fibers and determined the force-velocity relationships of rat, monkey, and human soleus type I

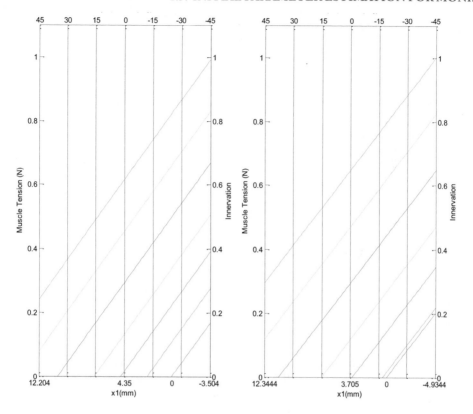

Figure 1.16: Length-tension curves generated by the linear model for monkey extraocular muscles (left). The innervations from upper to lower are 45N, 30N, 15N, 0, 15T, 30T, 45T (passive). On the right is length-tension curves of human extraocular muscle.

fiber (Widrick et al., 1997). Based on this study, we estimate that the monkey extraocular muscles contract faster than human extraocular muscles. Thus, B_1 of monkey is smaller than B_1 of human.

Since B_1 of human is 5.6 Ns/m, B_1 of monkey is estimated as 4 Ns/m since the peak velocity of human muscle is around 70% of the peak velocity of monkey. B_2 of monkey is estimated as 0.4 Ns/m. The linear muscle model's force-velocity relationships are shown in Fig. 1.17.

1.5.3   OCULOMOTOR PLANT PARAMETERS

K is first determined by steady-state analysis of the model,

$$\frac{180}{\pi r} K_{se}(F_{g0} - F_{t0}) = P_0 \theta \tag{1.70}$$
$$= (K_{st} K + 2K_{lt} K_{se})\theta$$

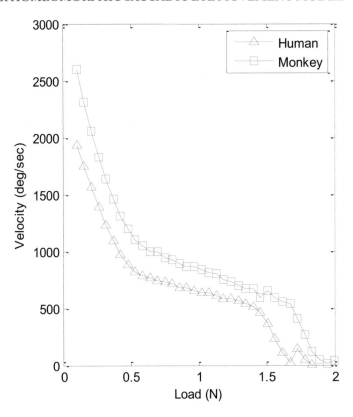

Figure 1.17: Force-velocity relationships for the model of monkey and human. (Human: $B_1 = 5.6\,$Ns/m and $B_2 = 0.5\,$Ns/m, Monkey: $B_1 = 4\,$Ns/m, $B_2 = 0.4\,$Ns/m.)

K should satisfy the steady-state of all eye positions, which yields a value of 10.21 N/m. B is then determined by assuming the dominant orbital time constant is 0.02 s, and

$$\frac{B}{K} = 0.02 \text{ s} \tag{1.71}$$

which yields $B = 0.204\,$Ns/m.

   J is the moment of inertia of the eyeball. Since the radius of monkey eye ball is smaller than the radius of human eye ball, and assuming that the densities are similar, the moment of inertia of monkey eye ball is estimated as 80% of the human eye ball, $J = 1.76 \times 10^{-3}\,$Ns^2/m.

   A comparison of the parameters estimated for monkey and the parameters of human is detailed in Table 1.3.

   Compared to the human muscle, the monkey muscle is stiffer, reflected by the larger elastic parameter K_{lt}. It also contracts faster than human muscle since it has smaller B_1. The firing fre-

Table 1.3: Comparison of parameters for monkey and human.

Parameter	Human	Rhesus Monkey
Radius of eye ball	11 mm (11.8mm in model)	10 mm
K_{se}	125 N/m	125 N/m
K_{lt}	60.7 N/m	77.66 N/m
B_1	5.6 Ns/m	4 Ns/m
B_2	0.5 Ns/m	0.4 Ns/m
F	$F = \begin{cases} 0.4 + 0.0175\lvert\theta\rvert, & \text{for } \theta > 0 \\ 0.4 - 0.0125\lvert\theta\rvert, & \text{for } \theta \leq 0 \end{cases}$	$F = \begin{cases} 0.55 + 0.0175\lvert\theta\rvert, & \text{for } \theta > 0 \\ 0.55 - 0.0125\lvert\theta\rvert, & \text{for } \theta \leq 0 \end{cases}$
K	16.34 N/m	10.21 N/m
B	0.327 Ns/m	0.204 Ns/m
J	2.2×10^{-3} Ns^2/m	1.76×10^{-3} Ns^2/m

quency of monkey motoneurons is faster than humans. The oculomotor plant parameters of monkey extraocular muscle are smaller than the ones of human because the monkey eye ball is smaller.

The transfer function for the oculomotor plant, based on Eq. (1.26), is

$$H(s) = \frac{\theta}{\Delta F} = \frac{\delta B_2 \left(s + \frac{K_{se}}{B_2}\right)}{s^3 + P_2 s^2 + P_1 s + P_0} \tag{1.72}$$

where $\Delta F = F_{ag} - F_{ant}$. Using the parameter values in Table 1.3, we have the transfer function for human as

$$H(s) = \frac{1.9406 \times 10^5 \, (s + 250)}{s^3 + 596 s^2 + 1.208 \times 10^5 s + 1.3569 \times 10^6} \tag{1.73}$$

and for monkey

$$H(s) = \frac{2.6904 \times 10^5 \, (s + 312.5)}{s^3 + 575.2 s^2 + 1.4829 \times 10^5 s + 2.7743 \times 10^6} \, . \tag{1.74}$$

There are three poles and the one zero in the transfer function shown in Eq. (1.72). Using the parameter values in Table 1.3 for human, the poles are

$$-292.22 + j168.63$$
$$-292.22 - j168.63$$
$$-11.92$$

and the zero is

$$250 \, .$$

For monkey, the poles are

$$-277.48 + j245.09$$
$$-277.48 - j245.09$$
$$-20.24$$

and the zero is

$$312.5 \, .$$

For human, the time constant for the real pole is 3.4 ms, and for the complex pole, 83.9 ms. Similarly, for monkey we have 3.6 and 49.4 ms.

Data and analysis of the monkey data is presented first in the next section. Following this, data and analysis of human data is presented, with inferences on neural control signals based on monkey.

1.6   MONKEY DATA AND RESULTS

Data[4] were collected from a rhesus monkey that executed a total of 27 saccades in our data set for 4°, 8°, 16°, and 20° target movements. Neuron data were recorded from the long lead burst neuron

---

[4]Details of the experiment and training are reported elsewhere [Sparks et al., 1976. Size and distribution of movement fields in the monkey superior colliculus, *Brain Research*, vol. 113, pp 21–34]. (Data provided personally by Dr. David Sparks.)

(5 saccades), excitatory burst neuron (17 saccades) and the agonist burst-tonic neuron (5 saccades). The firing of the burst tonic neuron is similar to the motoneuron that drives the agonist muscle during a saccade. Figure 1.18 shows the estimation results for three saccades (4°, 8°, and 15°). Initial estimates of monkey model parameters are listed in Table 1.3. The system identification technique provides close agreement between the displacement and velocity estimates from the model, and the displacement data and the derivative of the displacement data (velocity). The acceleration estimate from the model was the least accurate when compared with the second derivative of the displacement data (acceleration); it should be noted that the second derivative of the displacement data considerably amplifies the noise in the data. The accuracy of the results in Fig. 1.18 is typical for the other 4°, 8°, 16°, and 20° saccadic eye movements in our study.

Figure 1.19 shows the estimated neural inputs and active-state tensions that generate the saccades shown in Fig. 1.18. Also shown are the firing rates recorded from a single burst-tonic cell in a rhesus monkey (green line) for these saccades, scaled to match the height of N_{ag}. The shapes of model's neural inputs approximates the burst-tonic data during the pulse and slide very closely. The estimated agonist neural input N_{ag} clearly has similar shape as the firing rate data. It should be noted that the firing activity in the data comes from a single burst-tonic neuron. The neural input to the oculomotor plant is actually due to the firing of more than 1,000 motoneurons.

To evaluate the robustness of the estimation results, the neuron data are filtered by a discrete low-pass filter to obtain neuron-data-derived active-state tensions using Eq. (1.57). The filter uses the activation and deactivation muscle time constants estimated by the system identification previously described. A firing threshold is set to neglect the random and small firing before the bursting of the neuron. Figure 1.20 shows the data-derived active-state tensions. The neuron-data-derived active-state tensions are very close to those estimated by the system identification technique shown in Fig. 1.20. Using the data-derived active-state tensions as inputs to the 2009 oculomotor plant model, Fig. 1.18 also shows the generated eye position, velocity, and acceleration plotted against the data.

As with the results presented in Fig. 1.18, the data is closely approximated by the predictions of the model. It should be noted that no post-saccade phenomena (i.e., those with a dynamic overshoot or a glissade) are observed in any of the monkey data analyzed, and no PIRB in the antagonist muscle was detected by the system identification technique.

Important parameter estimates for all 27 saccades are shown in Figs. 1.21 and 1.22. The main-sequence diagram is shown in Fig. 1.21. Peak-velocity estimates follow an exponential shape as a function of saccade magnitude. Duration has a linear relationship with saccade magnitude for saccades above approximately 8°. For saccades less than 8°, duration is approximately constant with high variability. It should be noted that saccade duration is difficult to determine, especially for small saccades and may be a source of differences with other published data. The latent period is relatively independent of saccade magnitude.

The estimated agonist pulse magnitudes and durations are shown in Fig. 1.22 for all 27 saccades. The agonist pulse magnitude does not significantly increase as a function of saccade mag-

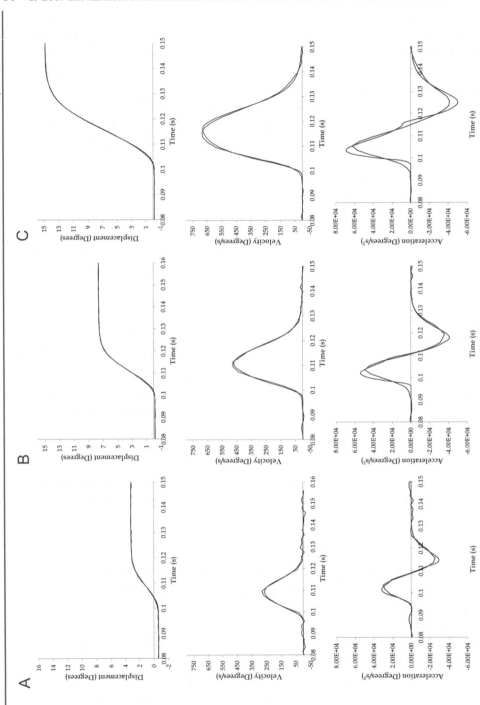

Figure 1.18: Eye position, velocity and acceleration for three different saccades in a rhesus monkey (A: 4 deg; B: 8 deg, and C: 15 deg). Blue lines are the model predictions and the red lines are the experimental data during the saccadic eye movement.

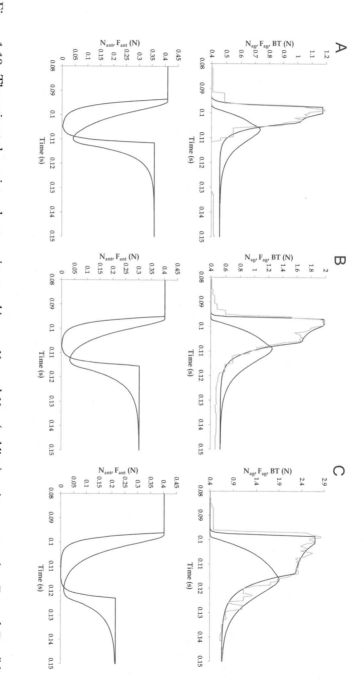

Figure 1.19: The estimated agonist and antagonist neural inputs N_{ag} and N_{ant} (red line), active–state tension F_{ag} and F_{ant} (blue line) for the three saccades (A: 4 deg, B: 8 deg, and C: 15 deg) shown in Fig. 1.18. Also shown are the firing rates recorded from a single burst-tonic cell in a rhesus monkey (green line) for these saccades, scaled to match the height of N_{ag}.

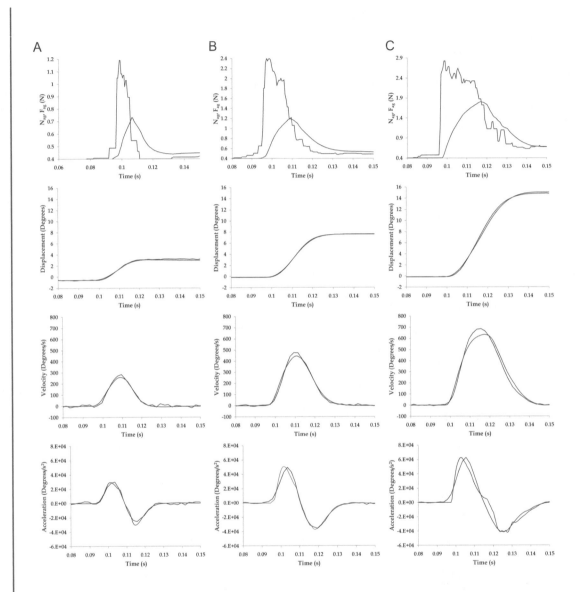

Figure 1.20: Active-state tensions (blue line) obtained from low-pass filtered firing rate of a burst-tonic neuron (red line) using the activation and deactivation time constants calculated in the parameter estimation. Also shown are the model predictions using the parameter estimates from the system identification technique for displacement, velocity, and acceleration and the data (red line is data and blue line is the model predictions). (A: 4 deg, B: 8 deg, and C: 15 deg.)

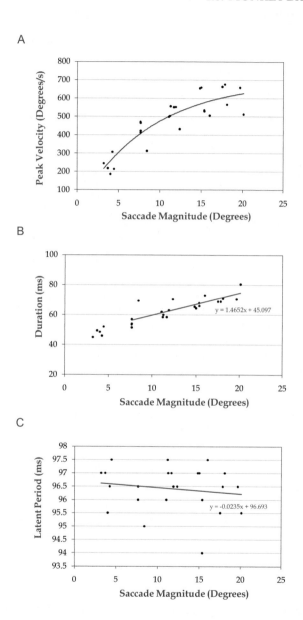

Figure 1.21: Main sequence diagram for monkey data. (A) Peak velocity vs. saccade magnitude. Fitted exponential curve to the data is $\dot{\theta}_{pv} = 694.65 \left(1 - e^{-0.1155\theta_{ss}} \right)$. (B) Duration vs. saccade magnitude with regression straight approximation for saccades larger than 5°. (C) Latent period vs. saccade magnitude based on the data.

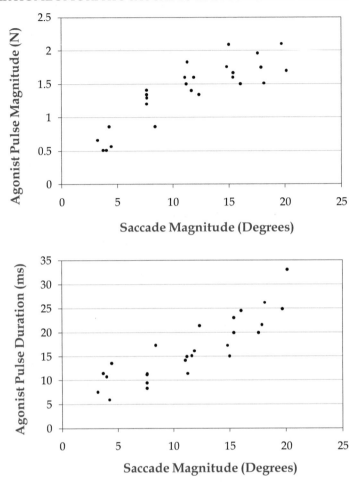

Figure 1.22: Agonist pulse magnitude (A) and duration (B) as functions of saccade magnitude for monkey data.

nitude for saccades larger than 8°, consistent with the time-optimal controller proposed by Enderle, J. (2002). For saccades under 8°, agonist pulse magnitude shows a linear increase in pulse magnitude vs. saccade magnitude, again in agreement with our theory for the saccade controller. A great variability is observed in the pulse magnitude estimates for saccades of the same magnitude, which is also observed by Hu et al. (2007) in their analysis of the firing rates in the monkey EBN. The agonist pulse duration increases as a function of saccade magnitude for saccades larger than 8°, and for smaller saccades, the duration of the agonist pulse is relatively constant as a function of saccade magnitude. Note that for all saccades, the pulse magnitude is tightly coordinated with the pulse duration. Interestingly, Fuchs and coworkers describe the firing rate for saccades greater than 10° as saturating, and for saccades up to 5°, they have the same approximate duration in monkey (Fuchs et al., 1985). We will discuss these details when the time optimal controller is presented.

1.7   HUMAN DATA AND RESULTS

Data[5] was collected from three human subjects executing 127 saccades, many with dynamic overshoots or glissades. Displayed in Fig. 1.23 are representative model estimates of saccades generated with a dynamic overshoot, a glissadic overshoot and normal characteristics. The model predictions for all saccades match displacement data and estimates of velocity very well, including saccades with a dynamic or a glissadic overshoot, with accuracy similar to those in Fig. 1.23. As noted before, the acceleration model prediction and the data-derived acceleration estimate were not as good as those of displacement and velocity.

The 8° saccade shown in Fig. 1.23 (A) of data and model predictions has dynamic overshoot. Note that the saccade with dynamic overshoot is caused by a PIRB firing in the antagonist neural input at approximately 220 ms. The PIRB induces prominent reverse peak velocity as shown.

Figure 1.23 (B) shows model predictions and data for a 8° saccade with glissadic overshoot. The glissade is caused by the PIRB in the antagonist neural input at approximately 223 ms. Notice the peak firing for a saccade with glissadic overshoot is smaller than one with dynamic overshoot. The PIRB induces reverse peak velocity that is smaller than the one with dynamic overshoot. In glissadic overshoot, the eye has an overshoot that returns to steady-state more gradually. As a result, the glissade has a smaller peak velocity.

A −12° normal saccade is shown in Fig. 1.23 (C). Normal saccades usually do not have a PIRB, although this is not absolute as the timing of the PIRB might offset the impact of the burst.

Important parameter estimates for all 127 saccades are shown in Figs. 1.24–1.27. The main-sequence diagram is shown in Fig. 1.24. Peak-velocity estimates from model are in close agreement with the data estimates of peak velocity and follow an exponential shape as a function of saccade magnitude. Duration has a linear relationship with saccade magnitude for saccades above 7°. For saccades between 3 to 7°, duration is approximately constant. It should be noted that saccade duration is difficult to determine, especially for small saccades and may be a source of differences with other published data. The latent period is relatively independent of saccade magnitude.

The estimated agonist pulse magnitudes and durations are shown in Fig. 1.25 for all 127 saccades. The agonist pulse magnitude does not significantly increase as a function of saccade magnitude for saccades larger than 7°, consistent with the time-optimal controller proposed by Enderle, J. (2002). For saccades under 7°, agonist pulse magnitude shows a linear increase in pulse magnitude vs. saccade magnitude, again in agreement with our theory for the saccade controller. A great variability is observed in the pulse magnitude estimates for saccades of the same magnitude, which is also observed by Hu et al. (2007) in their analysis of the firing rates in the monkey EBN. The agonist pulse duration increases as a function of saccade magnitude for saccades larger than 7°. For

[5]Details of the experiment are reported in Enderle and Wolfe (1988).

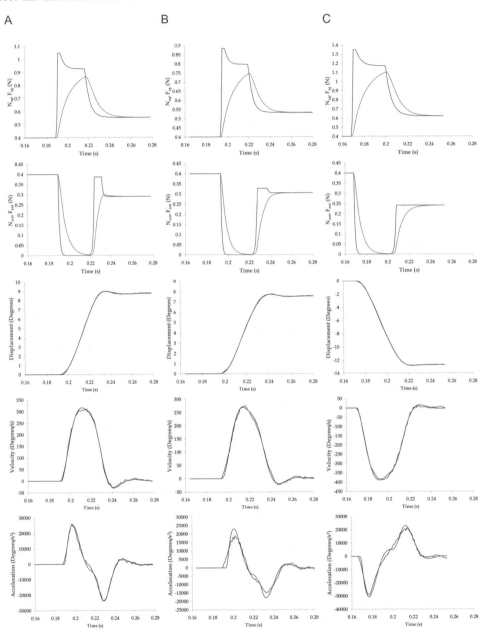

Figure 1.23: (A) Dynamic overshoot saccade of 8°; (B) glissadic overshoot saccade of 8°; and (C) normal −12° saccade. The first two lines of graphs are the active state tension and neural input calculated from the parameter estimation. Also shown are the model predictions using the parameter estimates from the system identification technique for displacement, velocity and acceleration and the data (red line is data and blue line is the model predictions).

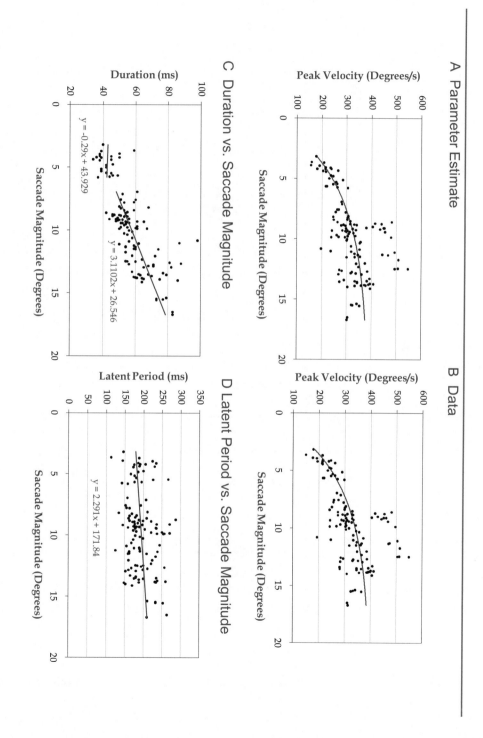

Figure 1.24: Main-sequence diagram for all 127 saccades from three human subjects. (A) Peak velocity vs. saccade magnitude from the model estimates, with regression fit $\hat{\theta}_{pv} = 390\,(1 - e^{-0.2\theta_{ss}})$. (B) Peak velocity vs. saccade magnitude from the data, with regression fit $\hat{\theta}_{pv} = 401\,(1 - e^{-0.2\theta_{ss}})$. (C) Duration vs. saccade magnitude based on the data. (D) Latent period vs. saccade magnitude based on the data. Note that the parameter estimation program did not update the duration or the latent period, thus a single graph for each is drawn.

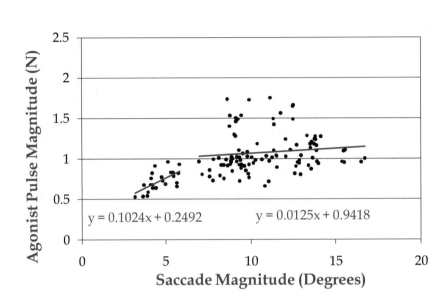

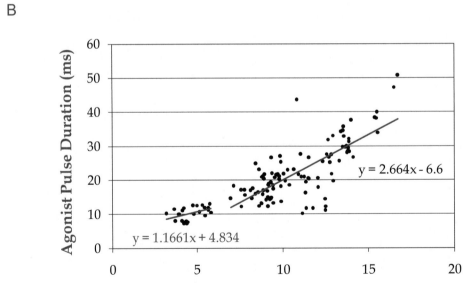

Figure 1.25: Agonist pulse magnitude (A) and duration (B) as functions of saccade magnitude.

saccades between 3 to 7°, the duration of the agonist pulse is relatively constant as a function of saccade magnitude. Note that for all saccades, the pulse magnitude is tightly coordinated with the pulse duration.

Figures 1.26 and 1.27 provide estimates for the PIRB in the antagonist motoneuron, where the PIRB induces a reverse peak velocity. Normal saccades usually do not have a notable rebound burst. Figure 1.26 describes the relationship between saccade magnitude and PIRB magnitude and duration. As shown, there are more saccades with dynamic overshoot in the abducting than adducting direction. There is also great randomness in the occurrence of PIRB. As shown in Fig. 1.26, the rebound burst magnitudes for dynamic overshoots are usually larger than the magnitudes for glissades. The rebound burst duration is approximately 12 ms, with considerable variation for saccades of the same size.

In Fig. 1.27, the antagonist onset delay is plotted. Normal saccades are clustered close to the origin, while moving further from the origin, glissades cluster in a band followed by saccades with a dynamic overshoot.

Shown in Fig. 1.28 are the estimates for the agonist and antagonist time constants. The agonist and antagonist activation time constants have a great impact on saccade dynamics. The agonist activation time constant showed great variability for saccades of the same size. No trends were observed between the agonist activation time constant and saccade magnitude. The other time constants showed less variability for saccades of the same size. Other parameter values had little variation around their initial estimates.

Figure 1.29 show the results of the system identification for estimating the parameters of the oculomotor plant. B_1 has the greatest movement from the initial parameter estimate, especially for positive eye movements. The other parameters have little variation about their initial values.

The objective of this chapter is to present a new model that more accurately characterizes horizontal saccadic eye movements using a third-order linear homeomorphic model of the oculomotor plant with a pulse-slide-step neural controller. Also included in the controller is PIRB in both the agonist and antagonist neural controller after marked inhibition. Of fundamental importance is the oculomotor plant used in the study of the saccade controller. If the oculomotor plant is not homeomorphic, then the controller obtained is suspect. Our model is robust in accurately simulating horizontal saccades that follow the main-sequence diagram, and a linear muscle model that is homeomorphic and has the nonlinear force-velocity and length-tension characteristics of muscle.

Model parameters are estimated using the system identification technique with both monkey and human data showing an excellent agreement between the model predictions and the data. Analysis of the monkey data gives support that we are able to accurately estimate agonist neural input during saccades, and by inference, to accurately estimate agonist neural input in humans. Another purpose of this chapter is to describe a time-optimal controller for horizontal saccades. Although discussed more completely in the next chapter, we discuss the some features of the neural network here in the context of PIRB and the time-optimal controller. It is thought that the EBN, located in the paramedian reticular formation (PPRF), monosynaptically projects to motoneurons

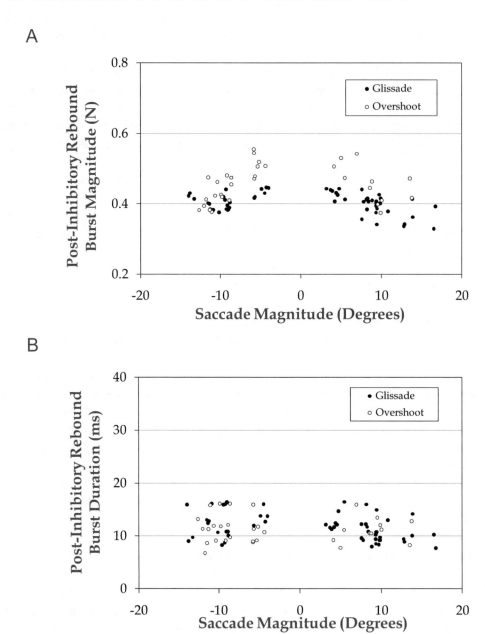

Figure 1.26: Post inhibitory rebound burst magnitude (A) and duration (B) as functions of saccade magnitude.

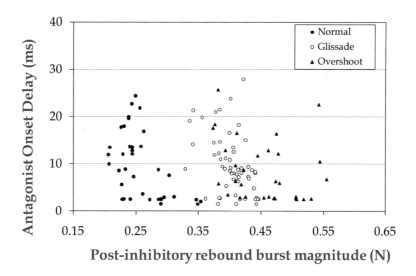

Figure 1.27: Post-saccade phenomena involving normal, glissade, and dynamic overshoot saccades.

(either the abducens or oculomotor), and that the discharge in the motoneuron during a saccade resembles a delayed EBN signal. The firing of the motoneurons is responsible for movement of the eyes. Another burst neuron in the PPRF, called the inhibitionary burst neuron (IBN), is also active during saccades with a discharge pattern similar to the EBN. IBN are involved with inhibiting the neuron sites that drive the antagonist muscle.

1.8 POST-INHIBITORY REBOUND BURST AND POST SACCADE PHENOMENA

Inhibition of antagonist burst neurons is postulated to cause an unplanned PIRB toward the end of a saccade that causes dynamic overshoots or glissades (Enderle, J., 2002). While some studies do not observe the rebound firing in the abducens neurons in monkeys (Fuchs and Luschei, 1970; Ling et al., 2007; Sylvestre and Cullen, 1999), PIRB are observed in the abducens motoneurons at the end of off-saccades in monkeys in other studies (for examples, see Robinson, D., 1981; Gisbergen et al., 1981). It has been noted earlier that saccades with dynamic overshoots or glissades do not occur with the same frequency in the monkey as in humans, and that they are absent from our monkey data.

Our theory is that, at least in humans, the antagonist PIRB causes a reverse peak velocity during dynamic overshoots or glissades in humans. The model predictions accurately match the velocity data for the entire saccade, including saccades with dynamic or glissadic overshoot. We were

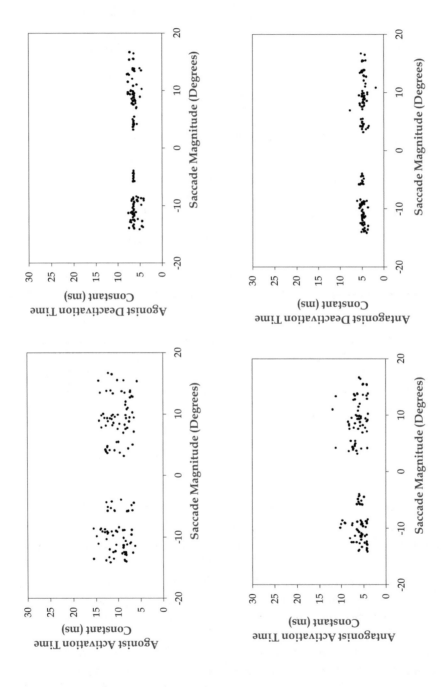

Figure 1.28:   Activation and deactivation time constants for the agonist and antagonist muscles.

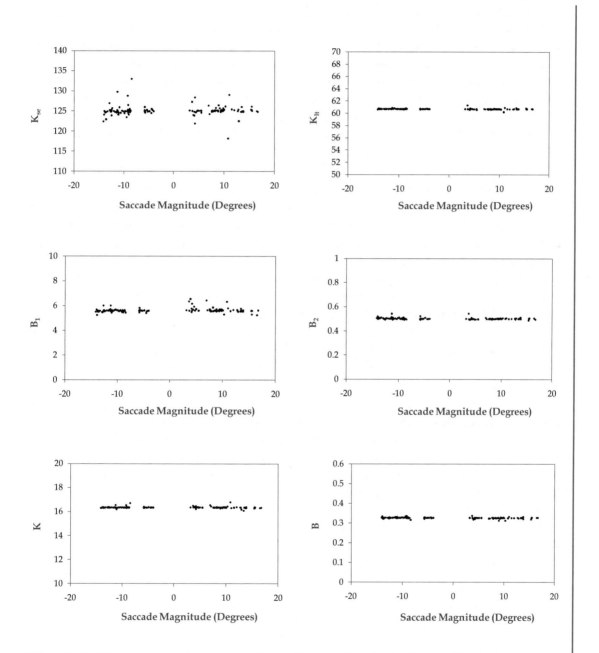

Figure 1.29: Final parameter estimates for the oculomotor plant obtained from 127 saccades.

unable to generate saccades with post-saccade behavior based on just the timing of the antagonist step, but we needed the PIRB to generate saccades with dynamic or glissadic overshoot.

Figures 1.26 and 1.27 summarize the characteristics of the 127 saccades collected from the three human subjects. The number of saccades with a glissade is larger than the number of normal saccades or those with a dynamic overshoot. Additionally, the incidence of dynamic overshoot decreases as saccade size increases. As shown, saccades with a dynamic overshoot typically have larger rebound burst magnitude than those with a glissade or with normal characteristics. The antagonist onset delay varies from 3 ms to approximately 25 ms. With a larger rebound burst, the onset delay is typically shorter for each type of saccade.

An inherent coordination error exists between the return to tonic firing levels in the abducens and oculomotor motoneurons during the completion of a saccade. During an abducting saccade, ipsilateral abducens motoneurons fire without inhibition and oculomotor motoneurons are inhibited during the pulse phase. Because the IBN inhibits antagonist motoneurons, resumption of tonic firing and PIRB activity in the motoneurons does not begin until shortly after the ipsilateral IBN's cease firing. This same delay exists in the abducens motoneurons for adducting saccades.

There are significantly more internuclear neurons between the contralateral EBN and the TN and the ipsilateral oculomotor motoneurons (antagonist neurons during an abducting saccade), than the ipsilateral EBN and TN and ipsilateral abducens motoneurons (antagonist neurons during an adducting saccade). Due to the greater number of internuclear neurons operating during an abducting saccade, a longer time delay exists before the resumption of activity in the oculomotor motoneurons after the pulse phase for abducting than adducting saccades.

Since the time delay before the resumption of activity in the oculomotor motoneurons after the pulse phase of a saccade is greater for abducting saccades than with adducting saccades, the incidence of saccades with dynamic overshoot should be greater for abducting saccades than adducting saccades. This is precisely what is observed in saccadic eye movement recordings; most saccades with dynamic overshoot occur in the abducting direction. Additionally, because the contralateral TN's firing rate decreases as ipsilateral saccade amplitude increases, the rate of dynamic overshoot decreases since fewer saccades have sufficiently high PIRB magnitudes. This is also what is observed in saccadic eye movement recordings.

Figure 1.30 illustrates the relationship between the PIRB magnitude and the antagonist onset time delay. As shown, it is possible for a normal saccade to have a small PIRB as long as the onset delay is small. As the onset delay increases, the PIRB must decrease or a saccade with dynamic or glissadic overshoot occurs.

1.9   TIME-OPTIMAL CONTROLLER

The general principle for a time-optimal controller for the horizontal saccade system is that the eyes reach their destination in minimum time that involves over 1,000 neurons. Each neuron contributes to the neural input to the oculomotor plant. In Section 3.6 of Book 1, we described the time-optimal control of saccadic eye movements with a single switch-time using a linear homeomorphic

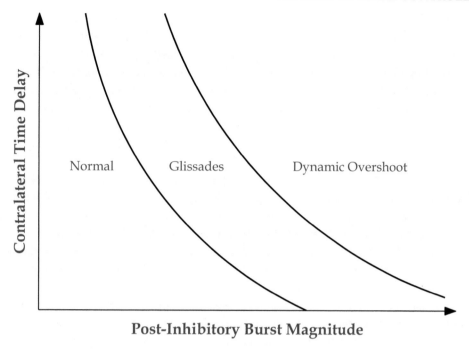

Figure 1.30: Relationship between antagonist onset time delay and the post-inhibitory rebound burst magnitude.

oculomotor plant for the lateral and medical rectus muscles. This section reexamines this earlier study using the updated oculomotor plant and a time-optimal controller constrained by a more realistic pulse-slide-step motoneuron stimulation of the agonist muscle with a pause and step in the motoneuron stimulation of the antagonist muscle, and physiological constraints.

The time-optimal controller proposed here has a firing rate in *individual* neurons that is maximal during the agonist pulse and independent of eye orientation, while the antagonist muscle is inhibited. We refer to maximal firing in the neuron as the intent of the system, which because of biophysical properties of the neuron membrane, slowly decay over time as described in the next chapter (Enderle, J., 2002). The type of time-optimal controller described here is more complex than the one previously described in Section 3.6 of Book 1 due to physiological considerations in the SC. The time-optimal controller operates in two modes, one for small saccades and one for large saccades.

The duration of small saccades have been reported as approximately constant (Fuchs et al. (1985) and reported here), and also as a function of saccade amplitude (e.g., Bahill et al., 1980). Estimating the saccade start and end time is quite difficult because it is contaminated by noise. We used a Kaiser filter to reduce the impact of noise, which others may not have implemented and possibly introduced a difference in results. Moreover, synchrony of firing will have a greater impact on the start time for small saccades than larger saccades since the beginning of the saccades is much more drawn out, making detection more difficult. In our analysis, a regression fit for the data is carried out in two intervals, one between 3 and 7°, and one for those greater than 7°. Our results indicate an approximately constant duration for small saccades and a duration that increases with saccade size for large saccades. Other investigators have used a single interval for the regression fit to a straight line, or a nonlinear function. It is possible that using the technique used here will result in a similar conclusion to ours. Since we did not analyze saccades less than 3°, judgment on saccade duration in this interval is delayed and supports future investigation.

We propose that there is a minimum time period that EBN's can be switched on and off, and that this is a physical constraint of the system. As shown in Fig. 1.24 (C), small saccades have approximately the same duration of 44 ms, and they do not significantly change as a function of saccade magnitude. Also, note that there is randomness in the response, where saccades with large pulse magnitudes are matched with shorter durations, and vice versa. As the saccade size increases for small saccades, we propose additional neurons are added to the agonist neural input up to 7°, where above this, all neurons are engaged.

In our model, we sum the input of all active motoneurons into the firing of a single neuron. Thus, as the magnitude of the saccades increases, the firing rate of the single neuron in our model increases up to 7°, after which it is maximal since all neurons are firing. Keep in mind, however, the firing rate of a real neuron is maximal and does not change as a function of saccade magnitude as easily seen in Fig. 4 in Robinson, D. (1981) and Fig. 2 in Gisbergen et al. (1981). The overall neural input for the agonist pulse is given by

$$N_{ag} = \begin{cases} N(\theta_T) \ N_{ag_i} & \theta < 7° \\ N_{ag_{max}} & \theta \geq 7° \end{cases} \tag{1.75}$$

where $N(\theta_T)$ is the number of neurons firing for a saccade of θ_T degrees, N_{ag_i} is the contribution from an individual neuron, and $N_{ag_{max}}$ is the combined input from all neurons. For small saccades, the commencement of firing of the individual neurons, or synchrony of firing, has a great impact on the overall neural input since the period of firing during the pulse is small (10 ms for the estimate in Fig. 1.25 (B)). Randomness in the start time among the active neurons means that the beginning of the saccade is more drawn out, than if they all started together. For smaller saccades, this may result in an incorrect start time, which then effects the duration. Any lack of synchrony can cause the overall agonist input to be smaller; this is a much larger factor for a small saccade than a large saccade since the pulse duration is much larger. It is very likely that during a saccade, neurons do not all commence firing at the same instant. This is seen in Fig. 1.25 (B) where there is a small slope to the regression fit.

Above 7°, the magnitude of the saccade is dependent on the duration of the agonist pulse with all neurons firing maximally. The agonist pulse magnitude as shown in Fig. 1.25 (A) is approximately a constant according to the regression fit. The duration of the agonist pulse increases as a function of saccade magnitude as shown in Fig. 1.25 (B).

The saccade controller described here is a time-optimal controller that differs from the one describe by Enderle and Wolfe (1987) because of the physiology of the system. Active neurons during the pulse phase of the saccade all fire maximally. For saccades greater than 7°, this is the same time-optimal controller described earlier by Enderle and Wolfe (1987). For saccades from 3 to 7°, the system is constrained by a minimum duration of the agonist pulse; saccade magnitude is dependent on the number of active neurons, all firing maximally, consistent with physiological evidence. In terms of control, it is far easier to operate the system for small saccades based on the number of active neurons firing maximally, rather than adjusting the firing rate for all neurons as a function of saccade magnitude as proposed by others. Thus, the system described here is still time-optimal based on physiological constraints.

Recently Harris and Wolpert (2006) described a saccade controller that optimizes speed and accuracy in support of a time-optimal controller, the same type of controller used by Enderle and Wolfe (1987). Harris and Wolpert (2006) 3-pole oculomotor plant was not homeomorphic as the plant described here, nor was their controller based on neuro-anatomical constraints or able to produce realistic saccades whether normal, or containing dynamic overshoots or glissades. Their numerator term in the oculomotor plant did not include a derivative term as suggested by our model in Eq. (1.26), which has a significant impact on the results. The main sequence diagram presented in their Fig. 2 does not have the characteristics of those reported elsewhere such as Bahill et al. (1980), with a leveling off of peak velocity at approximately 20 degrees. This could be due to the oculomotor plant used in their model.

Generally, saccades recorded for any size magnitude are extremely variable, with wide variations in the latent period, time to peak velocity, peak velocity and duration. Furthermore, this variability is well coordinated by the neural controller. Saccades with lower peak velocity are matched with longer saccade durations, and saccades with higher peak velocity are matched with shorter saccade durations. Thus, saccades driven to the same destination usually have different trajectories.

Hu et al. (2007) examined the variability in saccade amplitude, duration and velocity in the monkey by recording eye position and the EBN. To examine the reliability of the EBN, saccades with the similar amplitude and velocity were analyzed, and it was determined that the initial portion of the EBN firing rate had little variability, while the last portion of the burst had observable variability. The initial portion of the burst for a 10° and a 20° saccade shown in their Fig. 2 are approximately the same size and shape. The major difference between the 10° and a 20° saccade is that the 20° saccade had a longer burst duration. Furthermore, Hu and coworkers proposed that the activity in a single burst cell is not independent of, but strongly correlated with the activity of other burst neurons. To achieve the low variability in the EBN burst for the population, a low variability in the input or a special biophysical property of the burst neurons exists, or a combination of the two is proposed by Hu and coworkers.

CHAPTER 2

Neural Network for the Saccade Controller

2.1 INTRODUCTION

Saccades are characterized by a rapid shift of gaze from one point of fixation to another, eye movements that are used in reading and quick scanning. The purpose of the movement is to move the eyeball to the target quickly. Although the purpose is clear, the neural control strategy is not. Studies of the saccadic control mechanism have been based on the system identification technique and control systems, single-unit microelectrode recordings, muscle tension measurements, and general observations from the main sequence diagram. In comparison to other systems, the oculomotor system is the best understood of all human control systems. However, significant and important differences still exist regarding the control mechanism during saccadic eye movements. Physiological evidence indicates that saccades are controlled through a parallel-distributed network involving the cortex, cerebellum, and brain stem as shown in Fig. 2.1. In particular, the saccadic neural activity in the Superior Colliculus (SC) and the Fastigial Nucleus (FN) in the cerebellum have been identified as the saccade initiator and terminator, respectively, although neither is required for a saccade.

The saccade generator is discussed in this chapter on the basis of anatomical pathways and control theory. The literature is vast in this area (see Girard and Berthoz (2005); Krauzlis, R. (2005); Sparks, D. (2002); Scudder et al. (2002) for a review of models of the saccade system). As a result, the presentation here is not thorough, but rather a small sample from the literature. A list of abbreviations used for the neural sites involved with saccades is given in Table 2.1.

From each eye, the axons of the retinal ganglion cells exit and join other neurons to form the optic nerve. Then, the optic nerves from each eye join at the optic chiasm, where fibers from the nasal half of each retina cross to the opposite side. Axons in the optic tract synapse in the lateral geniculate nucleus (LGN), a thalamic relay, and continue to the visual cortex. This portion of the saccade neural network is concerned with the recognition of visual stimuli. Axons in the optic tract also synapse in the SC. This second portion of the saccade neural network is concerned with the location of visual targets and is primarily responsible for goal-directed saccades.

Under the time-optimal saccade controller described by Zhou et al. (2009) and Enderle, J. (2002), the saccade generator is initiated by the SC and terminated by the FN that operates under a time-optimal control strategy. To execute a saccade, a sequence of complex activities takes place within the brain, beginning with the detection of an error on the retina to the actual movement of the eyes to the destination. A saccade is directly caused by a burst discharge (pulse) from motoneurons

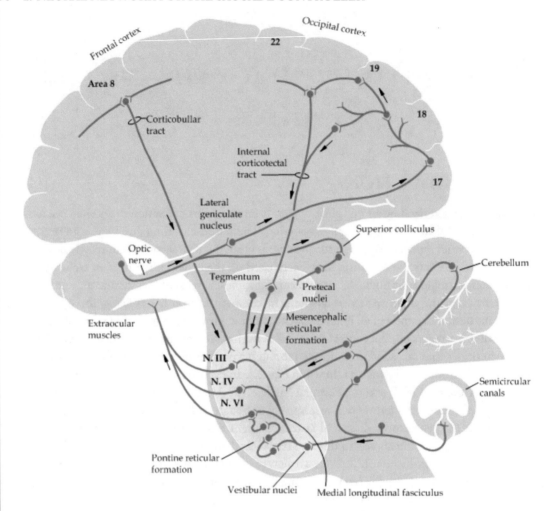

Figure 2.1: Diagram showing the major brain structures in the control of saccades.

stimulating the agonist muscle and a pause in firing from motoneurons stimulating the antagonist muscle. During periods of fixation, the motoneurons fire at a rate necessary to keep the eye stable (step). The model described here is based on Zhou et al. (2009). This model is a modification of that described by Enderle, J. (2002) with an input added to the Excitatory Burst Neurons (EBN) from the SC based on the work by Optican and coworkers (Ramat et al., 2005; Miura and Optican, 2006), and it still retains its input from the cerebellum as originally proposed. The saccade time-optimal controller is based on simplicity. Zhou et al. (2009) and Enderle, J. (2002) proposed that the cerebellum keeps track of the number of EBNs engaged in driving the agonist muscle (not the

Table 2.1: A List of Abbreviations.

Neural site	Abbreviation
Dynamic Motor Error	DME
Excitatory Burst Neuron	EBN
Fastigial Nucleus	FN
Inhibitory Burst Neuron	IBN
Long Lead Burst Neuron	LLBN
Medium Lead Burst Neuron	MLBN
Nucleus Reticularis Tegmenti Pontis	NRTP
Paramedian Pontine Reticular Formation	PPRF
Omnipause Neuron	OPN
Substantia Nigra	SN
Superior Colliculus	SC
Tonic Neuron	TN
Vestibular Nucleus	VN

individual EBN firing frequency), and the duration of the pulse to end the saccade. The difference in the individual firing rate in the EBN is not significant as compared to the number of EBNs firing. Moreover, monitoring and integrating the velocity feedback information from every neuron seems unnecessary.

2.2   NEURAL NETWORK

Regardless of the sensory input causing the fast eye movement, neural commands generate a saccade flow along a pathway called the final common pathway, which is illustrated in Fig. 2.2. The neurons within the final common pathway include the Medium Lead Burst Neuron (MLBN) in the Paramedian Pontine Reticular Formation (PPRF), Abducens Nucleus and the Oculomotor Nucleus. Although saccades can be stimulated from a variety of different sites, if the final common pathway is removed, no saccade will occur.

Under the time-optimal saccade controller described by Enderle, J. (2002) and Zhou et al. (2009), the saccade generator is initiated by the SC and terminated by the FN that operates under a time-optimal control strategy. To execute a saccade, a sequence of complex activities takes place within the brain, beginning from the detection of an error on the retina, to the actual eye movement. A saccade is directly caused by a burst discharge (pulse) from motoneurons, stimulating the agonist muscle and inhibiting the antagonist muscle. During periods of fixation, the motoneurons fire at a rate necessary to keep the eye stable (step). The pulse discharge in the motoneurons is caused by the EBN, and the step discharge is caused by the Tonic Neurons (TN) in the PPRF.

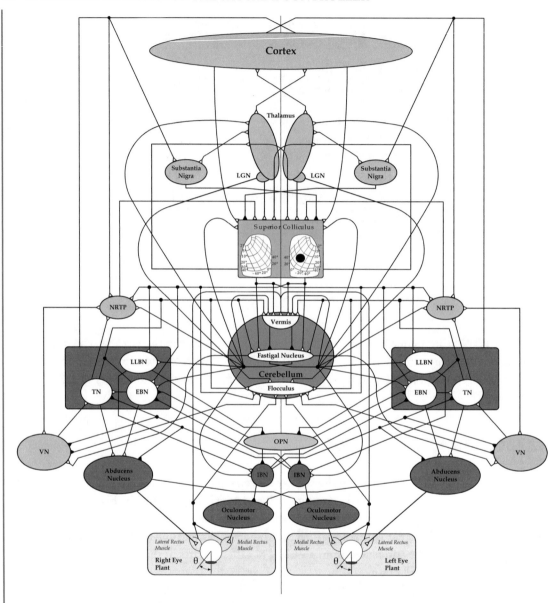

Figure 2.2: (A). Shown is a diagram illustrating important sites for the generation of a conjugate horizontal saccade in both eyes. Excitatory inputs are shown with △ inhibitory inputs are shown with a ▲. Consistent with current knowledge, the left and right structures of the neural circuit model are maintained. Since interest is in goal directed visual saccades, the cortex has not been partitioned into the frontal eye field and posterior eye field (striate, prestriate, and inferior parietal cortices).

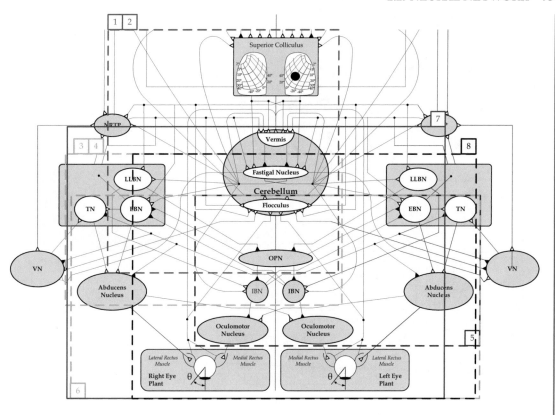

Figure 2.2: (B). Execution of a twenty degree saccade. Above is the part of the neural network illustrated in Fig. 2.2 (A) that generates the saccade, with boxes delineating the steps.

The model proposed here is a modification of that described by Enderle, J. (2002) with an input added to the EBN from the SC based on the work by Optican and coworkers, and still retains its input from the cerebellum as originally proposed (Ramat et al., 2005; Miura and Optican, 2006). The focus of the model presented here is primarily on the activity of the EBN and motoneurons, and thus details about the TN are limited in our presentation. The saccade time-optimal controller is based on simplicity. We propose that the cerebellum keeps track of the number of EBNs engaged in driving the agonist muscle (not the individual EBN firing frequency), and the duration of the pulse to end the saccade. The difference in the individual firing rate in the EBN is not significant as compared to the number of EBNs firing. Moreover, monitoring and integrating the velocity feedback information from every neuron seems unnecessary.

Consider the saccade network programmed to move the eyes 20° illustrated in Fig. 2.2 (A). Qualitatively, a saccade occurs according to the following sequence of events that are illustrated in Figs. 2.2 (B)–(H). Figure 2.2 (B) describes the 8 steps necessary to execute a saccade, with the

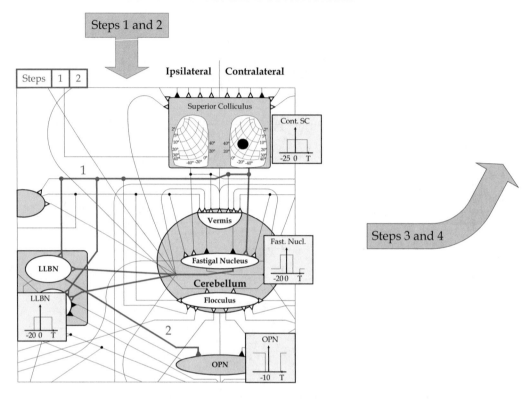

Figure 2.2: (C). 1) The contralateral SC stimulates the ipsilateral LLBN, EBN and the contralateral FN. 2) The LLBN then inhibits the tonic firing of the OPN.

remaining figures highlighting the neurons involved in each step. The timing and general shape of the firing in each neuron population is shown in the yellow blocks next to the neuron population. A more extensive discussion of the neuron sites active during a saccade is given later in this chapter.

1. The deep layers of the SC initiate a saccade based on the distance between the current position of the eye and the desired target as illustrated in Fig. 2.2 (C). The neural activity in the SC is organized into movement fields that are associated with the direction and saccade amplitude, and it does not involve the initial position of the eyeball whatsoever. Neurons active in the SC during this particular saccade in Fig. 2.2 are shown as the dark circle, representing the desired 20° eye movement. Active neurons in the deep layers of the SC generate an irregular high-frequency burst of activity that changes over time, beginning 18-20 ms before a saccade and ending sometime toward the end of the saccade; the exact timing for the end of the SC firing is quite random and can occur either before or after the saccade ends.

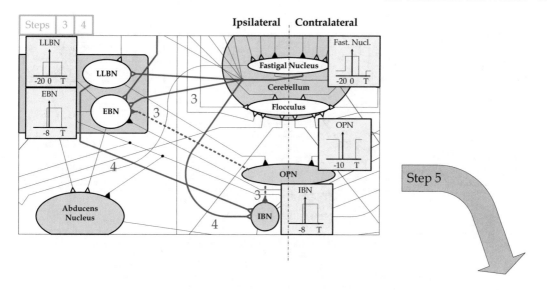

Figure 2.2: (D). 3) When the OPN cease firing (represented by dashed line 3 that is no longer active), the MLBN is released from inhibition and the IBN and the EBN are no longer inhibited. The ipsilateral EBN fire spontaneously given modest stimulation. The EBN are stimulated by the contralateral FN of the Cerebellum, and SC. However, FN stimulation is not required for a saccade to be generated. 4) The ipsilateral IBN are stimulated by the ipsilateral LLBN and the contralateral FN of the Cerebellum.

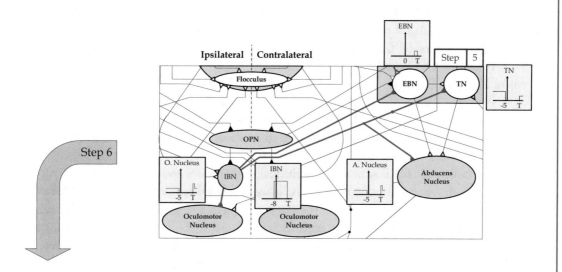

Figure 2.2: (E). 5) The burst firing in the ipsilateral IBN inhibit the contralateral EBN, TN, Abducens Nucleus, and the ipsilateral Oculomotor Nucleus.

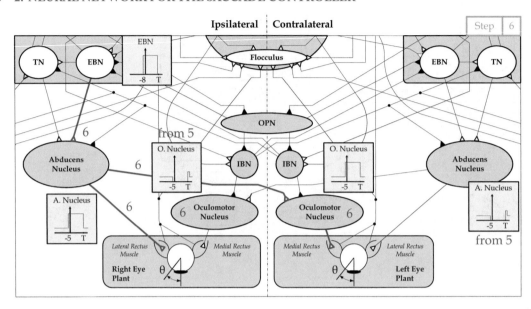

Figure 2.2: (F). 6) The burst firing in the ipsilateral EBN cause the burst in the ipsilateral Abducens Nucleus, which stimulates the ipsilateral lateral rectus muscle and the Contralateral Oculomotor Nucleus.

2. The ipsilateral LLBN and EBN are stimulated by the contralateral SC burst cells as shown in Fig. 2.2 (C). The LLBN then inhibits the tonic firing of the OPN. The contralateral FN also stimulates the ipsilateral LLBN and EBN.

3. When the OPN cease firing, the MLBN (EBN and IBN) is released from inhibition as shown in Fig. 2.2 (D). Some report that the ipsilateral EBN is probably not stimulated by the SC (Gandhi and Keller, 1997; Ramat et al., 2007). This conflict doesn't impact our model as we propose the stimulation of the EBN by other sites does not reflect the firing rate of the EBN, but that the EBN fire autonomously given weak stimulation. A mechanism for the EBN firing is presented later in this chapter.

4. The ipsilateral IBN is stimulated by the ipsilateral LLBN and the contralateral FN of the cerebellum. When released from inhibition, the ipsilateral EBN responds with a PIRB for a brief period of time. The EBN when stimulated by the contralateral FN (and perhaps the SC) enables a special membrane property that causes a high-frequency burst that decays slowly until inhibited by the contralateral IBN. The IBN may also have the same type of special membrane properties.

5. The burst firing in the ipsilateral IBN inhibit the contralateral EBN and Abducens Nucleus, and the ipsilateral Oculomotor Nucleus.

6. Shown in Fig. 2.2 (F) is the burst firing in the ipsilateral EBN that causes the burst in the ipsilateral Abducens Nucleus, that then stimulates the ipsilateral lateral rectus muscle and the

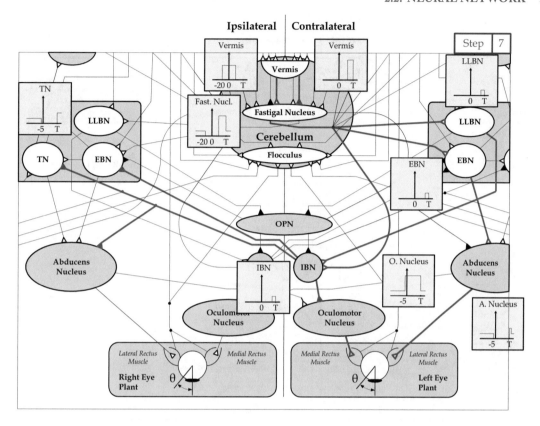

Figure 2.2: (G). At the termination time, the cerebellar vermis operating through the Purkinje cells inhibits the contralateral FN and stimulates the ipsilateral FN.

contralateral Oculomotor Nucleus. With the stimulation of the ipsilateral lateral rectus muscle by the ipsilateral Abducens Nucleus and the inhibition of the ipsilateral rectus muscle via the Oculomotor nucleus, a saccade occurs in the right eye. Simultaneously, the contralateral medial rectus muscle is stimulated by the Contralateral Oculomotor Nucleus, and with the inhibition of the contralateral lateral rectus muscle via the Abducens Nucleus, a saccade occurs in the left eye. Thus, the eyes move conjugately under the control of a single drive center.

7. At the termination time, the cerebellar vermis, operating through the Purkinje cells, inhibits the contralateral FN and stimulates the ipsilateral FN. Some of the stimulation of the ipsilateral LLBN and IBN is lost because of the inhibition of the contralateral FN.

The ipsilateral FN stimulates the contralateral LLBN, EBN, and IBN. Further simulation of the contralateral IBN occurs from the contralateral LLBN. The contralateral EBN then stimulates

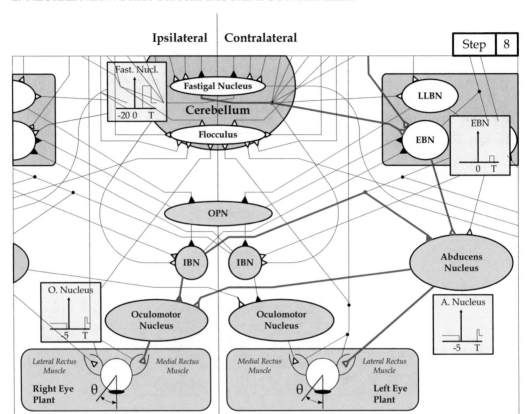

Figure 2.2: (H). The ipsilateral FN stimulation of the contralateral EBN allows for modest bursting in the contralateral EBN, which stimulates the contralateral Abducens Nucleus and ipsilateral Oculomotor Nucleus.

the contralateral Abducens Nucleus. The contralateral IBN then inhibits the ipsilateral EBN, TN, and Abducens Nucleus, and contralateral Oculomotor Nucleus.

With this inhibition, the stimulus to the agonist muscles ceases. In most saccades, the SC continues to fire even though the saccade has ended. This will be described later.

8. The ipsilateral FN stimulation of the contralateral EBN allows for modest bursting in the contralateral EBN (while still being inhibited by the ipsilateral IBN whose activity has been reduced) as shown in Fig. 2.2 (H). This then stimulates the contralateral Abducens Nucleus and ipsilateral Oculomotor Nucleus. With the stimulation from the contralateral EBN through the contralateral Abducens Nucleus and ipsilateral Oculomotor Nucleus, the antagonist muscles fire, causing the antagonist muscles to contract.

Once the SC ceases firing, the stimulus to the LLBN stops, allowing the resumption of OPN firing that inhibits the ipsilateral and contralateral MLBN and the saccade ends.

The cerebellum is included in the saccade generator as a gating element, using three active sites during a saccade: the oculomotor vermis, FN, and flocculus. The vermis is concerned with the absolute starting position of a saccade in the movement field and corrects control signals for initial eye position. Using proprioceptors in the oculomotor muscles and an internal eye position reference, the vermis is aware of the current position of the eye. The vermis is also aware of the signals (dynamic motor error) used to generate the saccade via the connection with the Nucleus Reticularis Tegmenti Pontis (NRTP) and the SC. The oculomotor vermis and FN are important in the control of saccade amplitude, and the flocculus, the perihypoglossal nuclei of the rostral medulla. The pontine and mesencephalic reticular formation are thought to form the integrator within the cerebellum. One important function of the flocculus may be to increase the time constant of the neural integrator for saccades starting at different locations from primary position.

The output of the FN is excitatory and projects ipsilaterally and contralaterally as shown in Fig. 2.2. During fixation, the FN fires tonically at low rates. Twenty milliseconds prior to a saccade, the contralateral FN bursts, and the ipsilateral FN pauses before discharging with a burst. The pause in ipsilateral firing is due to Purkinje cell input to the FN. The sequential organization of Purkinje cells along beams of parallel fibers suggests that the cerebellar cortex might function as a delay, producing a set of timed pulses, which could be used to program the duration of the saccade. If we consider nonprimary position saccades, different temporal and spatial schemes (via cerebellar control) are necessary to produce the same size saccade. The cerebellum acts as a gating device that precisely terminates a saccade based on the initial position of the eye in the orbit.

2.3 PARAMEDIAN PONTINE RETICULAR FORMATION

The PPRF has neurons responsible for the pulse and step discharges that drive the eyeball during a saccade. Neurons that fire at steady rates during fixation are called TN and are responsible for holding the eye steady. The TN firing rate depends on the position of the eye (presumably through a local integrator type network involving the EBN). The TN is thought to provide the step component to the motoneuron. There are two types of burst neurons in the PPRF called the LLBN and the MLBN. During periods of fixation, these neurons are silent. The LLBN burst at least 12 ms before a saccade and the MLBN burst less than 12 ms (typically 6-8 ms) before the saccade. The MLBN are connected monosynaptically with the Abducens Nucleus. Figures 2.3 and 2.4 shows the firing pattern for each of these neurons with the associated saccadic eye movement.

There are two types of neurons within the MLBN, the EBN and the IBN. The EBN and IBN label describes the synaptic activity on the other neurons. The EBN excites and is responsible for the burst firing. The IBN inhibits and is responsible for the pause. A mirror image of these neurons exists on both sides of the midline as shown in Fig. 2.2. The IBN inhibits the EBN on the contralateral side.

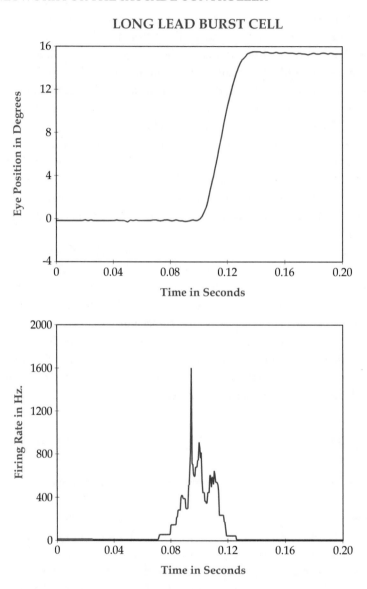

Figure 2.3: Extracellular single-unit recording from a (LLBN) and eye position obtained from a rhesus monkey during saccadic eye movements. Eye position data were recorded using magnetic coils; neural activity were recorded using tungsten microelectrodes. No filtering of the data was carried out, but the firing frequency was placed in the usual format of frequency of firing over 1 ms intervals, rather than the electrical activity itself. Details of the experiment and training are reported elsewhere [Sparks et al., 1976. Size and distribution of movement fields in the monkey superior colliculus, *Brain Research*, vol. 113, pp. 21–34.]. (Data provided by Dr. David Sparks.)

MEDIUM LEAD BURST CELL

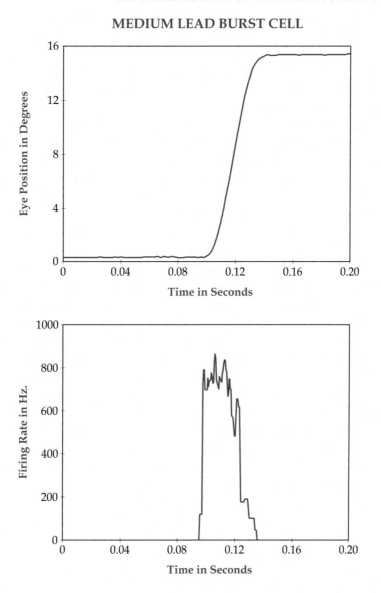

Figure 2.4: Extracellular single-unit recording from a MLBN and eye position obtained from a rhesus monkey during saccadic eye movements. Eye position data were recorded using magnetic coils, neural activity were recorded using tungsten microelectrodes. No filtering of the data was carried out, but the firing frequency was placed in the usual format of frequency of firing over 1 ms intervals, rather than the electrical activity itself. Details of the experiment and training are reported elsewhere [Sparks et al., 1976. Size and distribution of movement fields in the monkey superior colliculus, *Brain Research*, vol. 113, pp. 21–34.]. (Data provided by Dr. David Sparks.)

The EBN is located ipsilaterally just rostral to the Abducens Nucleus. They exhibit a burst of spikes for lateral saccades, and some produce monosynaptic excitation in the ipsilateral abducens motoneurons.

The IBN is located in the dorso-medial medullary reticular formation, just caudal to the Abducens Nucleus. They discharge as a burst of spikes for lateral saccades and produce monosynaptic IPSP in contralateral abducens motoneurons.

Also within the brainstem is another type of saccade neuron called the OPN. The OPN fire tonically at approximately 200 Hz during periods of fixation, and they are silent during saccades. The OPN stops firing approximately 10-12 ms before a saccade and resumes tonic firing approximately 10 ms before the end of the saccade. The OPN are known to inhibit the MLBN, and they are inhibited by the LLBN. The OPN activity is responsible for the precise timing between groups of neurons that causes a saccade. The existence of each of these PPRF neuron groups is uniformly accepted. However, the manner in which these neurons are connected is not uniformly accepted. Many models of the OPN use a bias input for its firing rate during periods of fixation.

Direct projections from the SC to the PPRF have been demonstrated anatomically. Previous electrophysiological experiments in the monkey have demonstrated direct projections from the SC to LLBNs exist, but they have failed to find definitive evidence for monosynaptic connections from the SC to EBN (Keller et al., 2000). In other experiments in the cat, direct connections to EBN were found (Chimoto et al., 1996). Keller conducted experiments to determine whether direct connections from the SC to EBN exist in the monkey. The experiment consisted of single-pulse stimuli delivered at sites in the SC at current levels well above those required to evoke saccades with pulse train stimuli shortly after the onset of ipsilateral or contralateral saccades and also slightly after the end of saccades. Out of the twenty-one recorded EBN's, none were activated by postsaccadic stimulation or during contralateral saccades. No evidence for direct connections to EBN was found in this study. The variance in results obtained for cat and monkey may be due to a species difference that reflects the more complex signal processing required in the monkey's saccadic system, or even inhibitionary projections from other neurons.

In contrast with the study of Keller and coworkers, existence of direct connections to the region of the EBN from both the SC and the FEF has been shown by anatomic means (Moschovakis et al., 1996; Olivier et al., 1993; Stanton et al., 1988; Harting, J., 1977). Nevertheless, stimulation of the deeper layers of the SC in the alert monkey by Raybourn and Keller (1977) were unable to activate EBN with single-pulse stimuli, although they did produce activation of EBN following triple-pulse stimuli with resultant latencies in the poly-synaptic range. Moreover, they described LLBNs that were also found in the PPRF in close proximity to EBN, which were readily activated from the SC, many in the range of latencies suggesting monosynaptic connections. LLBNs characteristics are not as clearly related to saccadic parameters such as duration and velocity as are those of EBN (Fuchs et al., 1985; Keller, E., 1991). Keller and coworkers noted that EBN in the monkey do not generally receive direct input from the SC, but they fall short of proving that no such connections exist because of difficulties in interpretation when stimuli are delivered during ipsilateral saccades (Keller et al., 2000). There is no clear direct evidence that LLBN project directly to the EBN (Moschovakis et al., 1996).

Perhaps, some of the difficulty in accepting that the EBN is not stimulated by the SC is discerning how the neurons fire with characteristics closely tied to the saccade characteristics. Without substantial stimulation, what causes the "spontaneous" firing in the EBN when it is released from inhibition during a saccade? How is the firing rate of the EBN controlled so that the accuracy of the saccade is maintained? Here, a model of the EBN is described using a Hodgkin-Huxley model of the neuron, where the threshold and time constant has been adjusted. With this updated biophysical property, the EBN is capable of firing at 1000 Hz automatically. With minor stimulation when released from inhibition, the EBN has tightly linked characteristics to the saccade as described in the main sequence diagram.

Variations in burst frequency, synchrony of firing and the number of motoneurons firing appear the most likely cause for the randomness observed from saccade to saccade. Here we propose that the major factor for saccade variability is the number of neurons firing in synchrony from saccade to saccade, and not variations in the firing rate. Sparks et al. (1976) describe the firing of neurons in the SC:

> "The precision or accuracy of a saccade results from the summation of the movement tendencies produced by the population of neurons rather than the discharge of a small number of finely tuned neurons. The contribution of each neuron to the direction and amplitude of the movement is relatively small. Consequently, the effects of variability or 'noise' in the discharge frequency of a particular neuron are reduced by averaging over many neurons. By reducing the effects of 'noise' in the discharge frequency of individual neurons, the large movement fields (which result in large populations of neurons being active during a specific movement) may contribute to, rather than detract from, the accuracy of the saccade."

As described later, the population of neurons in the SC are connected to neurons in the final common pathway that drive the eyes, and such statements about the SC neurons should also apply to the EBN.

Hu and coworkers report that the EBN burst is modulated by external factors such as attention and arousal level, and internal factors. They also point out that further investigation of the input to the EBN is needed. Moreover, their analysis does not support a velocity-based controller, such as the one proposed by Sylvestre and Cullen (1999), that is, when movements having similar amplitudes but different velocity profiles occur, it is not likely that this scenario represents a signal explicitly coding for particular saccade velocities.

Further, Thier and coworkers (2000) report a lack of support for a peak firing rate during the agonist pulse and the saccade velocity controller based on an individual neuron. They report that the mean discharge rate in the population of active neurons contributing to the agonist pulse is independent of saccade velocity.

Other saccade generator models, such as the Scudder model (1998) and the Gancarz and Grossberg (1998) model, are structured to provide a firing rate-saccade amplitude dependent signal. Cullen et al. (1996) used a system identification technique and found a firing-rate, saccade-amplitude-dependent controller. None of these studies used a homeomorphic ocular motor saccadic plant as is used here. However, the results shown here demonstrate that a first-order time-optimal controller is sufficient to generate saccades with random agonist pulse magnitudes that are tightly coupled with appropriate durations. Experimental data do not show a one-to-one relationship between EBN firing rate and saccade magnitude in our view. As demonstrated in this study, the agonist firing magnitude is independent of the size of the saccade for saccades larger than 7°, and only the pulse duration affects the size of the saccade. For saccades larger than 7°, all active neurons fire maximally and the duration of firing determines saccade magnitude. For saccades less than 7°, the agonist firing magnitude is determined by the number of active neurons.

A recent review of the saccade generator by Scudder and coworkers did not identify the excitatory inputs to the EBN with certainty (Scudder et al., 2002). Some have postulated the SC projecting directly to the EBN, others have the long lead burst neurons (LLBN) projecting directly to the EBN, and still others have both the SC and LLBN projecting directly to the EBN. In addition, others report that the FN also projects to the EBN, as well as the frontal eye fields and the central mesencephalic reticular formation. In 2006, Kaneko reports that the structure of the saccadic burst generator remains a mystery, and that saccades are largely ballistic and can be improved somewhat on line.

In 1995, Enderle and Engelken (Enderle and Engelken, 1995) proposed that the EBN, when released from profound inhibition, reacts with a PIRB and then continues to burst at a high-frequency with no or minimal input until just before the saccade ends. This type of input is a time-optimal input. Recent evidence by Miura and Optican has shown that when the OPN are lesioned, the EBN responds with a PIRB, but does not continue to fire afterwards (Miura and Optican, 2006). With this new evidence, the original EBN theory by Enderle and Engelken (1995) is easily amended to require an input that during or after the post-inhibitory burst, turns on the high-frequency burst process within the membrane (Zhou et al., 2009). Thus, when coming out of profound inhibition without an input to the EBN, a PIRB occurs in the EBN, but no high-frequency burst follows. During a saccade, the input to the EBN enables special membrane channels that cause the high-frequency burst during the pulse phase.

Further, we propose that these channels fatigue over time, resulting in a burst whose amplitude decays with time. The decay may be attributable to a reduction in ions, recruiting additional K^+ channels, accumulating inactivation of Na^+ channels, unknown channels, a depolarizing current stimulus, or energy necessary to maintain the peak firing rate.

It is clear that the discharge in the inputs to the EBN do not resemble the discharge in the EBN, so it is unlikely that the EBN is firing as a typical neuron in response to the input. There is great variability in firing of those neurons proposed as inputs to the EBN but much less variability in the EBN discharge with similar profiles as described by Hu et al. (2007).

Therefore, another mechanism must be involved, which, when turned on by a sufficiently sized input, initiates the high-frequency burst in the EBN. Fuchs and coworkers describe the firing rate for saccades greater than 10° as saturating (Fuchs et al., 1985). Van Gisbergen and coworkers reported that 64% of the EBNs had an invariant relationship between motor error and firing rate (Gisbergen et al., 1981). Also evident in the data presented by Gisbergen et al. (1981), and Robinson, D. (1981) is that the PIRB occurs at the beginning of the burst that drives the agonist muscle and at the resumption of firing for the antagonist muscle toward the end of the saccade. The data shown in Fig. 1.19 is consistent with those reported by Gisberen et al. (1981) and others. Additional studies of the membrane properties of the EBN are necessary to further explore the saccade generator.

Although it has been proposed that the PIRB that stimulates the antagonist muscle at the end of the saccade acts as a break, we propose that it is unplanned and random, and is also accounted for by the saccade generator. Kapoula et al. (1986) reported that not all monkeys have PIRB, and that may be the reason that some investigators have not observed them in the monkey (Ling et al., 2007; Sylvestre and Cullen, 1999). We did not observe any PIRB in the monkey data analyzed in this study. It is very common in humans as observed in our data and those by others.

2.3.1   HODGKIN-HUXLEY MODEL OF AN EBN

For completeness, the 1995 model of the EBN is described here. To investigate the effect of threshold voltage on the firing characteristics of a neuron, simulations are presented using the nonlinear Hodgkin-Huxley model (Hodgkin et al., 1952) of neuron described by the circuit diagram in Figure 2.5. This model, originally described by Enderle and Engelken (1995) and Enderle, J. (2002), describes the membrane potential at the axon hillock due to conductance changes. The following equation is the node equation for the circuit, and it is parameterized for the squid giant axon defining the membrane potential V_m as a function of stimulus current I_m and active gate conductance for sodium and potassium.

$$I_m = \bar{g}_K\, n^4\, (V_m - E_K) + \bar{g}_{Na}\, m^3 h\, (V_m - E_{Na}) + \frac{(V_m - E_l)}{R_l} + C_m \frac{dV_m}{dt} \qquad (2.1)$$

where

$$\frac{dn}{dt} = \alpha_n\,(1 - n) - \beta_n n$$
$$\frac{dm}{dt} = \alpha_m\,(1 - m) - \beta_m m$$
$$\frac{dh}{dt} = \alpha_h\,(1 - h) - \beta_h h$$
$$\alpha_n = 0.01 \times \frac{V + 10}{e^{\left(\frac{V+10}{10}\right)} - 1} \quad ms^{-1}$$

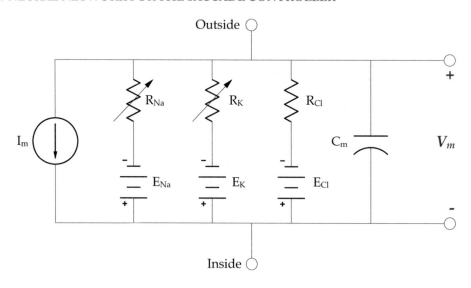

Figure 2.5: Circuit model of an unmyelinated section of squid giant axon. The channels for K^+ and Na^+ are represented using variable voltage-time conductances given by $R_K = \frac{1}{\bar{g}_K n^4}$ and $R_{Na} = \frac{1}{\bar{g}_{Na} m^3 h}$. The passive gates for Na^+, K^+, and Cl^- are given by a leakage channel with resistance, $R_l = \frac{1}{0.3 \times 10^{-3}}\Omega$. The batteries are given by the Nernst potential for each ion, $E_l = 49.4V$, $E_{Na} = 55V$ and $E_K = 72V$.

$$\beta_n = 0.125 e^{\left(\frac{V}{80}\right)} \quad ms^{-1}$$

$$\alpha_m = 0.1 \times \frac{V + 25}{e^{\left(\frac{V+25}{10}\right)} - 1} \quad ms^{-1}$$

$$\beta_m = 4 e^{\left(\frac{V}{18}\right)} \quad ms^{-1}$$

$$\alpha_h = 0.07 e^{\left(\frac{V}{20}\right)} \quad ms^{-1}$$

$$\beta_h = \frac{1}{e^{\left(\frac{V+30}{10}\right)} + 1} \quad ms^{-1}$$

$$V = V_{rp} - V_m \quad mV$$

$$\bar{g}_K = 36 \times 10^{-3} S$$

$$\bar{g}_{Na} = 120 \times 10^{-3} S$$

To investigate the effects of low threshold voltage, simulations using the Hodgkin-Huxley model of a neuron are presented using SIMULINK. Threshold voltage is defined as when the sodium current, I_{Na}, characterized by the m and h differential equations, is greater than the potassium current, I_K, characterized by the n differential equation, and leakage current. Since sodium current changes more quickly than the potassium current at the beginning of the action potential, changes in threshold voltage are accomplished by changing a parameter value in the sodium equation. Since sodium current has a rising component (the m differential equation) and a falling component (the h differential equation), threshold is modified here using the m differential equation. Thus, changes to threshold are carried out by changing the value of 25 to 10 in the algebraic equation for α_m, yielding

$$\alpha_m = 0.1 \times \frac{V + 10}{e^{\left(\frac{V+10}{10}\right)} - 1} \quad \text{ms}^{-1} . \tag{2.2}$$

The value of 10 is selected since threshold is approximately $V_m = -45\text{mV}$, resting potential $V_{rp} = -60$ mV, and the variable $V = V_{rp} - V_m = -60 + 45 = 15$ mV. Thus, to change threshold to -60 mV, 15 is subtracted from the quantity ($V + 25$) leaving ($V + 10$) in the m differential equation. To change the firing rate of the Hodgkin-Huxley neuron model so that it bursts at 1000 Hz, the right-hand side of the n, m and h equations are multiplied by 35,000, effectively reducing the apparent time constant for these differential equations to match the data.

To illustrate how changing threshold voltage affects the firing rates of the neuron, three simulations are presented in the following figures. Figure 2.6 illustrates a normal action potential stimulated by a current pulse of 20 μA for 4 ms with no initial hyperpolarization. Figure 2.7 illustrates a single burst firing for a neuron after coming out of marked hyperpolarization with the normal Hodgkin-Huxley model. Note this is without stimulation. Unique in this simulation is that no excitatory stimulus is used to cause the action potential. An action potential is caused by the hyperpolarization. Further, after the refractory period is over, an action potential is elicited by stimulating the neuron with a current pulse of 20 μA for 4 ms after the refractory period (not shown).

Figure 2.8 illustrates spontaneous burst firing without stimulation in a neuron after coming out of marked hyperpolarization at 20 ms using the modified Hodgkin-Huxley Eq. (2.2) in 7.3.1, an action potential occurring each time V_m reaches -60 mV. This continues without pause until the membrane is hyperpolarized at 80 ms. At this point, when the membrane returns to a constant voltage of approximately -80mV. This model does not have a stable resting potential and does not need to be depolarized to spontaneously fire. In order to match EBN firing rate as in Fig. 4 of Robinson, D. (1981), the right-hand side of the n, m, and h equations are multiplied by 35,000 from 20-30 ms to achieve a 1000 Hz firing rate, and then by 30,000 from 30-80 ms to achieve an approximate 800 Hz firing rate. Further, the current pulse is reapplied over a 20 ms interval from 60-80 ms to match the decay from steady-state firing observed in the data.

After release from hyperpolarization, the simulation shown in Fig. 2.9 has a firing rate of approximately 1000 Hz initially, drops to 800 Hz, and then returns to zero after the neuron is hyperpolarized again. Notice that instead of having a plateau firing rate of 800 Hz in Fig. 2.10, a slow decay in the interval 30-80 ms is modeled by lowering the constant from 30,000 linearly over the interval. Figure 2.11 summarizes the firing frequency in Fig. 2.10, with the EBN firing rate calculated from the inverse of the time interval between each action potential. A duration of 60 ms

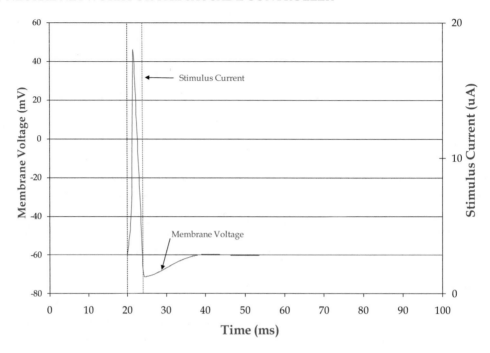

Figure 2.6: An action potential simulated with the Hodgkin-Huxley model. The current pulse starts at 20 ms with a magnitude of 20 μA and then turns off at 24 ms.

EBN burst usually results in a 10° saccade. The results in Fig. 2.10 summarized in Fig. 2.11 matches the characteristics of EBN data shown in Fig. 4 of Robinson, D. (1981) very closely (see figure for the 11° saccade).

If a depolarizing current stimulus is applied as the membrane comes out of hyperpolarization with this model, the delay between the release of the membrane to the first action potential is reduced, as compared to Fig. 2.8. Moreover, the time between action potentials is reduced while the depolarizing current stimulus is applied, resulting in a higher firing rate. This is shown in Figs. 2.10 and 2.11.

Shown in Fig. 2.12 is extracellular single-unit recordings from within the Abducens Nucleus, eye position and velocity obtained from a rhesus monkey during saccadic eye movements of 4, 8, 12, 16, and 20 degrees. While this set of data do not fire at the 1000 Hz, previously described in Figure 4 of Robinson, D. (1981) and the simulations, the EBN neuron model described here is easily modified to give the same type of firing activity by changing the multiplying constant in the n, m, and h equations.

Presented in this section is a mechanism based on biophysical phenomenon using a Hodgkin-Huxley model of neuron to simulate observed EBN burst firing during saccades. This behavior is

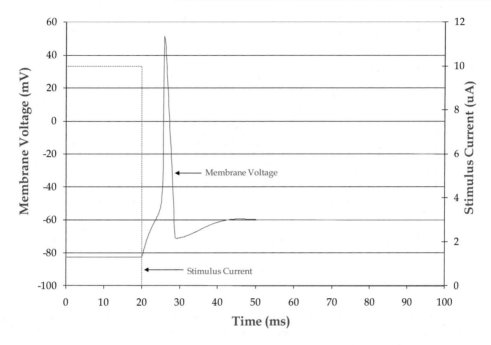

Figure 2.7: With the Hodgkin-Huxley model of Eq. (2.1), a −10 mA current pulse is used to hyper-polarize the cell membrane to approximately −80 mv for a very long time, and then turns off at 20 ms. After release from hyperpolarization at 20 ms, a single action potential occurs, without any stimulation.

described by neural firing rates of 800-1000 Hz that drives the eyes to their destination without any regard for the size of the saccade. While other neuron models could have been used, the Hodgkin-Huxley model provides a simple and straightforward mechanism that captures the essence of a fast firing neuron such as the EBN. Clearly, the linkage of threshold, via the sodium conductance channel provides an easy method of creating an automatically firing neuron. In fact, there may be other mechanisms that cause this phenomenon as described next.

Direct projections from the SC to the PPRF have been demonstrated anatomically in numer-ous studies. Previous electrophysiological experiments in monkeys have demonstrated that direct projections from the SC to LLBN exist, but recently they have failed to find evidence for monosy-naptic connections from the SC to EBN. Other investigators support direct connections between the SC and EBN. Perhaps, some of the trouble in accepting that the EBN is not directly stimulated by the SC or other neurons is the difficulty in discerning how the EBN firing rate characteristics are so closely tied to the saccade without direct stimulation by a neural controller. Since a direct cause and effect relationship is normally expected in such a system, just releasing a group of neurons from inhibition that causes "spontaneous" firing does not appear reasonable at first. Another difficulty is

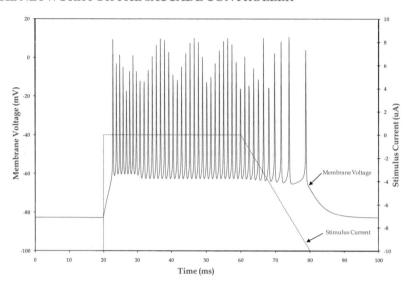

Figure 2.8: The Hodgkin-Huxley model of Eq. (2.1), with the equation for α_m replaced by Eq. (2.2) and the time constant for the n, m, and h differential equations scaled appropriately is used in this simulation. A $-10\,\mu A$ current pulse is used to hyperpolarize the cell membrane to -80 mv for a very long time, and then is turned off at 20 ms, and reapplied at 60 ms. After release from hyperpolarization at 20 ms, the neuron fires spontaneously at approximately 1000 Hz without any stimulation until it is hyperpolarized again at 60 ms.

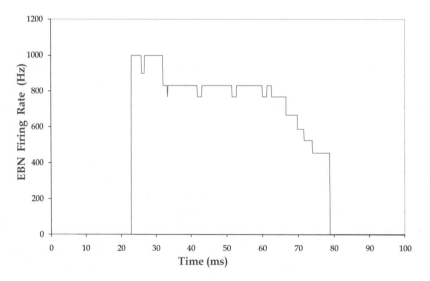

Figure 2.9: EBN firing rate calculated from the previous figure. The firing rate is calculated from the inverse of the time interval between each action potential.

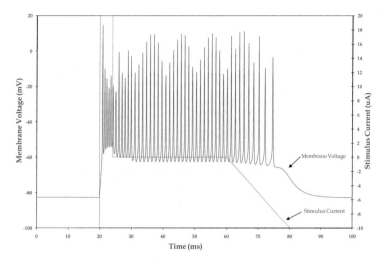

Figure 2.10: The Hodgkin-Huxley model of Eq. (2.1), with the equation for α_m replaced by Eq. (2.2) and the time constant for the n, m, and h differential equations scaled appropriately is used in this simulation. A -10 μA current pulse is used to hyperpolarize the cell membrane to -80 mv for a very long time, and then is turned off at 20 ms, and reapplied at 60 ms. At 20 ms, a pulse of 20 μA is applied for 4 ms. After release from hyperpolarization at 20 ms and because of a depolarizing stimulus current, the neuron fires at very high levels. After the depolarizing stimulus current is removed, the neuron fires spontaneously at approximately 800 Hz without any stimulation, until it is hyperpolarized again at 60 ms.

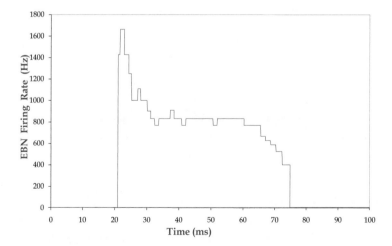

Figure 2.11: EBN firing rate calculated from Fig. 2.10. The firing rate is calculated from the inverse of the time interval between each action potential.

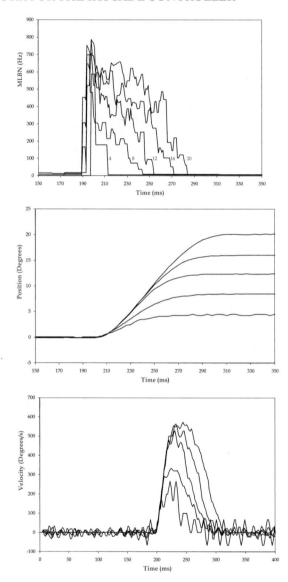

Figure 2.12: Extracellular single-unit recordings from within the MLBN, eye position and velocity obtained from rhesus monkeys during saccadic eye movements. Eye position data were recorded using magnetic coils. Neural activity was recorded using tungsten microelectrodes. No filtering of the data was carried out, but the firing frequency was placed in the usual format of frequency of firing over 1 ms intervals, rather than the electrical activity itself. Details of the experiment and training are reported elsewhere [Sparks et al., 1976. Size and distribution of movement fields in the monkey superior colliculus, *Brain Research*, vol. 113, pp. 21–34.]. (Data provided by Dr. David Sparks.)

the relationship between the firing rate of the EBN and saccade accuracy since the two are tightly coupled. In this section, a model of the EBN is described using a Hodgkin-Huxley model of the neuron in which the threshold and time constant have been adjusted to yield a spontaneously firing neuron that has EBN characteristics. With this updated biophysical property in the Hodgkin-Huxley model, the EBN is capable of firing at 1000 Hz automatically and without stimulation when released from inhibition. It has tightly linked characteristics to the saccade based solely on the duration of the burst firing under time optimal control (Enderle and Wolfe, 1987).

It should be noted that the EBN have been reported to receive projections from the vestibular fibers and burster-driving neurons in the vestibular nuclei, sites that are not active during goal directed saccades (Galiana, H., 1991; Ohki et al., 1988; Markham, C., 1981). It is not clear how these inputs would impact the EBN, but further study is certainly warranted.

The role of the LLBN is not exactly clear based on experimental evidence, with some supporting an LLBN to EBN projection, and some is not supporting the LLBN to EBN projection (Moschovakis et al., 1996; Gancarz and Grossberg, 1998). It is clear that the LLBN receives projections from the SC, and the LLBN may be involved in an integrating mechanism, but the integrative mechanism is not uniformly accepted (Gancarz and Grossberg, 1998). The firing characteristics of the LLBN follow those of the SC rather closely, but they do not follow the EBN after the OPN have been inhibited. It is clear that the firing characteristics of the LLBN are not tightly related to the characteristics of the saccade (main sequence), even during the time interval when the OPN is inhibited. The EBN has been described with

(1) A resettable leaky integrator in models of the EBN to match the characteristics of experimental data in some studies (Moschovakis et al., 1996; Gancarz and Grossberg, 1998).

(2) A BIAS input in order to match the characteristics of experimental data for small saccades (for example, see Gancarz and Grossberg, 1998).

Neither (1) nor (2) above are required with the EBN model proposed here. There is no memory associated with the spontaneously firing EBN, and the neuron fires at high rates for all saccades, which matches the experimental data.

The focus of the work in this section has been the EBN, but it can also be applied to the OPN, which does not seem to have any inputs as well. Some have proposed that the OPN are stimulated by a thalamic circuit, while others seem to suggest an unknown BIAS circuit. In fact, the neuron model presented here can be easily parameterized to match the OPN firing rates of 100-200 Hz.

2.3.2   COMPONENTS OF THE BURST

Shown in Fig. 2.13 are sketches of the EBN firing rate drawn to generally match the data in Fig. 4 of Robinson, D. (1981) with (A) and the data in Fig. 5(a) of Gancarz and Grossberg (1998) with (B). The interval 0 to T_1 is the minimum duration for EBN burst firing as supported by physiological evidence. For example, a 1° saccade would have a burst of T_1s, as would a 4° saccade. This interval represents the time it takes to switch off and on the OPN due to the inherent time delays in signal

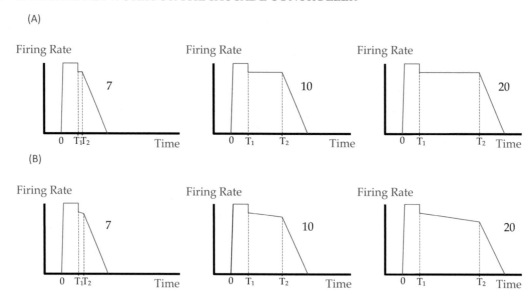

Figure 2.13: Block sketch of EBN firing rates for 7°, 10°, and 20° saccades drawn to match the data in Robinson, D., 1981 shown in (A), and in Gancarz and Grossberg, 1998 in (B). Notice that the functions are identical from 0 to T_1 for all cases, where this represents the minimum time for an EBN burst.

transmission. The interval T_1 to T_2 is the time that the EBN drives the eyes to their destination. The interval T_2 until the EBN ceases firing is when the OPN resumes its inhibition of the EBN. This part of the waveform is the same for saccades of all sizes.

The EBN model presented here is designed to have a constant plateau of firing during the interval 30 to 60 ms to model the firing rate of Fig. 4 of Robinson, D. (1981). It can also be modified to have a small reduction in firing rate during this interval to match experimental data shown in Fig. 5(a) of Gancarz and Grossberg (1998). The decay can be implemented as previously described with a linear reduction in firing rate of this interval, and it might be attributable to a reduction in ions or energy necessary to maintain the firing rate. As noted previously, the firing rate of the EBN is quite random from saccade to saccade. Gancarz and Grossberg indicate that the decreasing firing rate in the interval T_1 to T_2 for the EBN is due to IBN inhibition of the LLBN. Since the LLBN in their model drive the EBN, a reduction the LLBN firing rate causes a reduction in the EBN firing rate. Such a strategy implies that the EBN reflects an error signal. Such results are not consistently observed in the data published in the literature.

A minimum duration for an EBN neuron once released from inhibition is T_1 as shown in the Fig. 2.13. T_1 is the smallest possible interval in which the EBN can be switched on and off. This phenomenon is observed in the data, even for microsaccades. To account for minimum burst duration, fewer neurons in the SC fire for small saccades, as shown Fig. 2.14. This figure shows a

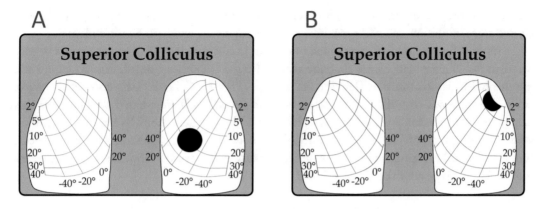

Figure 2.14: A detailed view of the retinotopic mapping over of the superior colliculus for a (A) 20° movement and a (B) 2°. Notice the locus of points for the 2° movement is smaller than that for the 20° movement implies that fewer neurons are firing for the smaller movement. The movement fields within the superior colliculus also reflects the number of neurons firing for saccades less than 7° are fewer than those firing for saccades greater than 7°. For saccades above 7°, the movement field is approximately constant.

detailed view of the SC for a 20° movement and a 2° movement. Notice the locus of points for the 2° movement is smaller than that for the 20° movement implies that fewer neurons are firing for the smaller movement. Above 7°, the size of the movement field remains constant in the SC. Below 7°, the size of the movement field decreases as it approaches zero. Since fewer cells firing in the SC means fewer cells in the LLBN, fewer EBN cells are released from inhibition via OPN release, and fewer cells are driving the eyes to their destination. This explains why small saccades all have approximately the same duration up to about 7°. We will discuss this further in Section 2.4.

2.3.3   POST INHIBITORY REBOUND BURST FIRING

Electrophysiological evidence for PIRB firing activity during saccadic eye movements is prevalent in the literature (as noted previously). This behavior is described by high neural firing rates without any regard for the size of the eye movement with minor stimulation at the start of EBN activity during the first 10 ms of firing. Post inhibitory rebound burst firing begins with a rapid rise to a peak occurring within a few ms, and then decaying to a lower steady-state firing rate at approximately 10 ms. The decay may be attributable to a reduction in ions, recruiting additional K^+ channels, a depolarizing current stimulus, accumulating inactivation of Na^+ channels, or energy necessary to maintain the peak firing rate. Furthermore, PIRB firing activity after marked hyperpolarization is postulated to occur in the EBN due to a low membrane threshold voltage within the axon hillock of the neuron. With this biophysical property, a single neuron is capable of automatically firing at high rates with minor stimulation when released from inhibition. According to this hypothesis, when released from inhibition, the EBN maximally fire automatically and with low stimulation.

Support for post inhibitory rebound burst firing activity is derived from the reports by many investigators. Jahnsen and Llinas describe rebound burst responses from thalamic neurons after very marked hyperpolarizations (Jahnsen and Llinas, 1984a,b). Other investigators report neurons within the thalamus that allow them to serve as single cell oscillators based on multi-threshold activity levels (Contreras et al., 1997; Destexhe et al., 1996). Because the EBN has no known inputs besides the FN, and possibly the SC, the EBN are modeled as firing spontaneously when released from inhibition. This appears to be an essential property of the neuron, since EBN fire after release from inhibition even after the FN is lesioned.

2.4   SUPERIOR COLLICULUS[1]

The SC is divided into two lobes, one on either side of the midsagittal line, as shown in Fig. 2.14, and in relation to other brain sites in Fig. 2.3. Each lobe has an architecture representative of the contralateral, retinotopic field. Neural signals sent to the SC are from a direct connection with the retina through the optic nerve and indirectly from the visual cortex. Each lobe of the SC relates to the opposite half of the visual field. Specific input/output connections for the SC in the generation of goal directed saccades are illustrated in Fig. 2.2.

Seven anatomical layers make up the SC, which have been traditionally considered two functional regions, the superficial layer, and the intermediate and deep layers. The superficial layer generally receives the visual input, and the deep layers receive input from the somatic sensory and auditory systems. The first three layers make up the superficial layer and the remaining four are divided evenly between the intermediate and deep layers. The superficial region is considered the visual layer, where information is received concerning activity in the visual field. Little interaction occurs between the superficial and deeper layers (Sparks, D., 1986). The intermediate and deep layers are involved with generation of the desired motor command, though there are differences between these two subgroups which will be addressed shortly (Raybourn and Keller, 1977; Albano and Wurtz, 1981; Sparks and Nelson, 1987; Lee et al., 1988).

The SC receives information about the desired target location from several different parallel redundant networks to initiate a saccade. The redundancy in this network allows for the integration of inputs from several sources and provides safeguards against damage to any of the pathways (Wurtz and Goldberg, 1972). The entire SC is organized to form a retinotopic mapping across each layer such that polar coordinates in the visual field are mapped to Cartesian coordinates along each layer, as shown in Fig. 2.14 (Robinson, D., 1972; Ottes et al., 1986). Nakahara et al. (2006) describe the activity in the SC as autonomous due to intrinsic connections without eye movement feedback.

---

[1]Some of the material in this section is based on the previously published paper: Short and Enderle (2001), A Model of the Internal Control System Within the Superior Colliculus, Biomed. Sci. Instru., 37:349–354, and the Master's Thesis by Steven Short, A Model of the Internal Control System Within the Superior Colliculus, University of Connecticut, December 2000.

Activity in certain loci within the SC corresponds to either the current location of a point of interest or the desired position to which the eyes should move. Cells in the deep and intermediate layers of the SC discharge strongly for a saccade of a certain amplitude and direction, but they also fire weakly for saccades within a certain range from this optimal position. Shown in Fig. 2.3 is the firing rate for an LLBN during a saccade. The firing rate of the SC neuron looks similar to the LLBN. Thus, during generation of a saccade, a certain region of cells will be firing for that saccade while the rest will be silent. Saccades of any amplitude larger than 7° typically have a fixed area of activated cells, while for saccades of smaller amplitude, the area of activated cells becomes smaller as 0 is approached (Sparks, D., 1986).

The superficial layers of the SC receive projections from several sources. Cells from the retinal ganglion transmit information through the optic nerve to the ipsilateral SC from the contralateral visual field on the retina. This information is then relayed out through the pulvinar nucleus of the thalamus (Fig. 2.15) primarily to a broad area of the cerebral cortex (Sparks, D., 1986). Projections

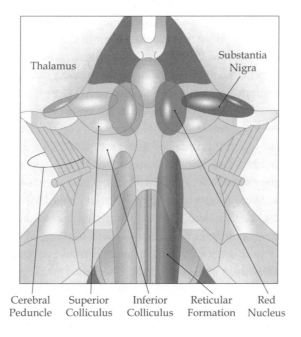

Figure 2.15: Diagram shows a frontal view of the SC and other brain structures, as well as some of the connections between them.

are also sent to the pretectal nuclei and the lateral posterior nuclei of the thalamus (Ito, M., 1984). The thalamus is a known relay center in the brain and transmits this information to several places, including the visual or striate cortex. After processing, a modified form of this signal is then sent back to the superficial layer, presumably as a source of comparison. Projections from the superficial layer to the intermediate and deeper layers have been found, but there is considerable question as to what the exact purpose of these projections is (Sparks, D., 1986).

The intermediate and deep layers of the SC receive projections from several different sources. Some of the inputs affect both layers. These are described first, followed by descriptions of signals uniquely important to one layer or the other. Both layers receive signals via projections from the prestriate, middle temporal, and posterior parietal cortices (PPC). The PPC is involved with visual attention. The lateral intraparietal area (LIP) of the PPC indicates intended amplitude and direction for targeting purposes. The middle temporal cortex is concerned with detection of motion in the visual field and the direction of motion. Visual information sent to the striate cortex from the superficial layer, as described previously, is relayed after modification back to the intermediate and deeper layers of the SC and thus completes an indirect pathway from superficial to deep layers. The caudal portion of the FN of the cerebellum projects to the contralateral SC at the level of both layers and is involved in controlling the accuracy of the saccades initiated by the SC (Robinson, D., 1981; Ito, M., 1984; Keller, E., 1989). Both intermediate and deep layers send excitatory signals out to the contralateral PPRF in the pons, as described in the previous section.

There are a number of projections that are important in the deeper intermediate layers of the SC. The nucleus prepositus hypoglossi (NPH) projects to the deep SC (Stechison et al., 1985). The role of the NPH is to monitor neural activity relating to the ongoing saccade. Thus, the projection to the SC may play a role providing gaze error feedback to the SC (Leigh and Zee, 1999). The central mesencephalic reticular formation (cMRF), located in the mesencephalon, projects to the deeper layers of the SC and receives input from the LLBN of the SC. The cMRF has a topographic organization similar to the SC and is active primarily for horizontal saccades. Projections to the SC from the cMRF, termed reticulo-tectal long lead burst neurons or RLLB, appear to be qualified to supply the SC with an efference copy signal relating the metrics of the current saccade (Moschovakis et al., 1996). Interestingly, similar neurons have been found in lower phyla, indicating that this may be part of the oldest aspects of the saccadic control system (Moschovakis et al., 1996). The deeper intermediate layer also receives a projection from the perihypoglossal nucleus, which is a motor nucleus that controls head and possibly neck muscles during combined head-eye gaze movements. Additionally, saccades can be generated from auditory signals (Engelken and Stevens, 1989; Engelken et al., 1991a; LaCroix et al., 1990). Projections out of the SC deep layers are sent to the raphe nucleus interpositus (RIP) in the contralateral PPRF and the OPN. Axons that terminate on the OPN have been found to originate from cells not only at the rostral pole but also over one third of the more rostral locations. There have been both inhibitory and excitatory connections found and there is some question as to the organization and function of these projections (Buttner-Ennever et al., 1999; Ohtsuka and Nagasaka, 1999).

The intermediate burst layer receives projections from the movement-related neurons of the frontal eye fields (FEF). These neurons fire for all saccades regardless of whether they are visually guided or not. The FEF is capable of eliciting saccades independently of the SC, but information as to the intended location of interest is shared between these two neural centers (Bruce et al., 1985;

Bruce and Goldberg, 1985). Projections from this layer also send signals to the FEF. The signal sent to the SC by the FEF is also sent to the caudate nucleus (Goldberg and Bushnell, 1981). Excitation causes the caudate nucleus to inhibit the substantia nigra pars reticulata (SN). The SN normally provides tonic inhibition to the intermediate layer of the SC. Inhibition of the SN releases the SC from inhibition and allows it to burst, generating the desired position signal (Graybiel, A., 1978). The FEF signals the generation of saccades for several different types of saccades, such as memory-guided or visual, but it is not heavily involved in the generation of spontaneous saccades, as the SC is. Thus, the SN does not demonstrate consistent pauses in inhibitionary action for spontaneous saccades (Hikosaka and Wurtz, 1983a). Instead, the zona incerta, a nucleus of the subthalamus area, provides an inhibition signal to the rostral pole of the deeper intermediate layers. The activity of the zona incerta pauses for all saccades (Ricardo, J., 1981; Hikosaka and Wurtz, 1983b; Moschovakis et al., 1996). If the FEF is lesioned such that no signal is sent to the caudate nucleus to silence the SN, saccades can still be made and the inhibition of the SN can be overcome (Kandel et al., 2000). Efferent projections of the intermediate layers terminate at the ipsilateral FN, as well as the contralateral cerebellar vermis and nucleus reticularis tegmenti pontis (NRTP).

The anatomical projections just presented involve the interaction of the SC with other brain structures. There are also a number of important anatomical connections that have been found within the layers of the SC and between the two lobes. Neurons in the intermediate layers of the SC exhibit very little activity for the majority of time that the eyes are fixated on a target. During a brief interval before a saccade starts, a region of cells within the intermediate layer begins to burst maximally to signal the intended amplitude and direction of the desired saccade (Sparks and Hartwich-Young, 1989). Because of this behavior, these cells have been termed burst cells (Sparks, D., 1978; Munoz and Wurtz, 1995b). The majority of cells within the deeper layers of the SC exhibit a different behavior. As signals enter the deep layers, activity increases for about one hundred milliseconds until a saccade is initiated (Sparks et al., 1976). However, this activity is not of maximal frequency, and it includes a greater percentage of the cells in these layers than the active area of the burst cells for any one saccade (Munoz and Wurtz, 1995a,b). The center of highest activity in these cells is not as clearly delineated as it is in the burst layer. The activity of these cells increases to the time when the burst cells fire and when those cells at the center of the deep layer activity burst as well. The frequency of the burst is not quite as high as in the burst layer, but it is much higher than the prior exhibited frequency. These cells have been named buildup cells (Sparks, D., 1978; Munoz and Wurtz, 1995b).

A minority of the cells within the deeper layers located at the rostral pole of this region do not follow this buildup pattern. Instead, these cells are active during the time prior to activation of the buildup neurons. The level of activity in these cells declines as the buildup neurons become excited so that they are negligibly active when the burst neurons fire. These cells are believed to be part of an internal control system within the SC whose purpose is to suppress unwanted saccadic activity. Thus, they are called fixation cells (Munoz and Wurtz, 1992, 1993a,b). The last cell type within the SC is the interneuron, whose purpose is to relay inhibitionary or excitatory signals from one locus to another within the SC. It is important to stress that the delineation between layers is not exact, and thus it is not possible to describe in millimeters where the buildup layer ends and burst layer begins. This is also true of the end of the buildup neurons and beginning of the fixation neurons. As a result, neurons that exhibit buildup cell characteristics have been found in regions commonly thought to contain only burst neurons and cells with movement fields for small angle (< 2°). Saccades have been found in the rostral pole, considered the location of fixation neurons (Munoz and Wurtz, 1995b).

2.4.1   SC SEQUENCE OF ACTIVITY IN THE GENERATION OF A SACCADE

The sequence of events that generate a saccade within the SC begins with the appearance of a target of interest in the visual field. The information relating position on the retina passes through the superficial layers of the SC and is relayed to the LGN. The processed information returns via the LGN and reaches the deeper layers of the SC. Any additional inputs to the deeper layer that relate a desired eye movement, as described previously, are received as well. Activation of individual neurons within the deeper layers causes activation of surrounding neurons within the movement field of that cell. As many neurons become excited and excite cells nearby them, a net area of excitation results with a locus of highest activity within this area of buildup cells (Horwitz and Newsome, 1999). This system facilitates target selection, as the final eye movement amplitude and direction selected are the net average of the inputs or the amplitude and direction for those cells firing with the greatest frequency in the deeper layers.

The buildup cells become active about 100 ms before saccade initiation and remain active at less than 200 Hz up until about 25 ms before saccade initiation (Munoz and Wurtz, 1995a). Signals are sent via interneurons to the burst neurons in the intermediate layers to activate cells whose movement field is centered at the desired amplitude and direction (Behan and Kime, 1996; Meredith and Ramoa, 1998; Munoz and Istvan, 1998). However, these cells are unable to fire because the burst layer is inhibited by the influence of the fixation neurons in the rostral pole. These cells project, via interneurons, inhibitory signals to the rest of the intermediate layers of the SC to ensure that no signals are sent to the eyes to move, thus keeping the eyes fixated on the target (Munoz and Istvan, 1998). The fixation neurons also send excitatory signals to the OPN in the RIP to help keep the eyes fixated (Buttner-Ennever et al., 1999; Ohtsuka and Nagasaka, 1999). However, the buildup cells also send inhibitory signals, through excitation of inhibitory interneurons, to the fixation neurons. As buildup activity increases, the fixation pole's activity decreases.

In the absence of any influences outside the SC, the pause in fixation cell activity results in release of burst cells from inhibition. Then, the burst cells excite the LLBN to move the eyes to the desired location. However, both the substantia nigra and the zona incerta exert inhibitory influences on the intermediate and deeper layers of the SC, though their effects on the buildup layer do not completely inhibit the buildup layer. The substantia nigra pauses for visually directed saccades, while the zona incerta pauses for all saccades. Thus, activation or inhibition of the zona incerta and the substantia nigra will externally control the activity of the SC. It has been shown that the SC can overcome the influence of the substantia nigra, which makes sense since not all eye movements are visually directed. The exact mechanism of zona incerta control is unknown, but the cerebellum may be involved to fine-tune saccade metrics. Therefore, the activity of burst neurons, and to a degree, the buildup neurons can be prematurely silenced. If the substantia nigra and zona incerta are lesioned, the saccade still ends with the normal error with respect to target location via cerebellar control.

According to our model, the number of active neurons during the pulse phase of the saccade needs to be a function of saccade size from 3 to 7°. Logically, this function could fall to the SC. Sparks and coworkers described increasing movement field of activity in the SC up to 10°, which then remained constant for those greater than 10° (Sparks et al., 1976); such a mechanism could meet our needs. Current theory, however, points to a constant field of activation involving 30% of the SC area (Anderson et al., 1998; Munoz and Wurtz, 1995a,b). Munoz and Wurtz presented data on the SC closed movement field that had cells closest to the 0 with the smallest movement field, which increased as saccade size increased in their Fig. 10A, which may indicate a smaller movement field for smaller saccades (Munoz and Wurtz, 1995b).

Anderson and coworkers reported a relationship between the firing rate in buildup cells and saccade size, which then affects the number of fixation cells and OPNs that are turned off, and the number of EBN allowed to fire (Anderson et al., 1998). Further, Krauzlis reports that many neurons in the rostral SC modulate their firing rates during small saccades (Krauzlis, R., 2005). This process is adequate to generate the mechanism described in this paper. In another publication, Das and coworkers describe weights representing synaptic strength from burst cells that increased by a factor of three from the rostral (smallest) to the caudal (largest) colliculus (Das et al., 1995; Anderson et al., 1998), thus providing another possible mechanism. While the exact mechanism of SC behavior during saccades is not well understood, perhaps the findings of this chapter might be helpful in delineating its role.

2.4.2   SUPERIOR COLLICULUS MODEL OF THE MOVING HILL

The mechanism that enables the SC to function without the influence of outside controls is the hypothesized moving hill process. This hypothesis states that the sequence of events within the SC begin, as described previously, with the buildup layer becoming excited and influencing excitation of the burst layer. The burst layer fires about 25 ms before the eyes actually begin to move (Munoz and Wurtz, 1995b). Once the saccade begins, the locus of activity in the buildup layer moves across the retinotopic map in a net caudal to rostral direction. The cMRF projects the desired eye displacement back to the buildup layer. This causes an increase in firing rate at a location rostral to the current locus of activity, thus moving the locus of activity in a rostral direction. As the saccade nears its end, the error goes to zero and the "hill" of activity moves to the rostral pole (Munoz et al., 1991; Munoz and Wurtz, 1995a). This activates the fixation neurons. In turn, they inhibit the burst layer to hold the eyes at the new location. This is best seen in the experiment by Munoz and Wurtz (1995a) where continuous excitation of a caudal burst layer location elicited a series of eye movements. The eyes move the distance defined by the optimum activated movement field and then the fixation neurons resume firing. Thus, there is an inherent control system for the generation and of saccades by the SC. The movement has not been clearly observed to reach the rostral pole for all saccades in all experiments. Indeed, researchers have found that saccades can be partially or significantly clipped, with eye movement ending before activity has ceased within the SC. As will be discussed in the next section, control of the end of the saccade is not entirely dependent on the drive from the SC, and it is controlled by the cerebellum.

Very few saccade generator models incorporate a functional model of the SC and the influence of the cerebellum on the SC. Recent work on the intracollicular activities within the SC during a saccade described a moving hill of neural activity (Munoz and Wurtz, 1995a,b, 1993a,b, 1992; Munoz and Istvan, 1998; Munoz et al., 1991). However, Anderson et al. (1998) observed very little movement of activity within the SC, therefore, questioning the existence of a moving hill of activity. Further work by Port et al. (2000) demonstrated a moving hill within the SC. A model of the SC is presented here based on the hypothesis of Munoz and coworkers, which explains all of the observations found by Munoz, Port and Anderson and coworkers (Short and Enderle, 2001). This model generates only horizontal movements and focuses on the series of events within the SC using SIMULINK to simulate saccades, but it could easily be expanded to horizontal and vertical saccades.

Activity occurs in up to three-quarters of the SC map in the buildup layer, the actual amount activated dependent on size of saccade. Several studies have been done to examine the population activity both spatially and temporally. Both Munoz and Wurtz (1995a,b), and Anderson et al. (1998) support activity in the buildup layer encompasses a larger portion of the whole map for any size saccade than activity for the same saccade in the burst layer. Moreover, the size of the active zone changes with the amplitude of the saccade. However, Munoz and Wurtz demonstrated a movement of the locus of activity within the buildup layer as a saccade develops so that the active region in the buildup layer moves rostrally across the retinotopic map and excites the fixation neurons, ending the saccade. Anderson and coworkers observed some although there was some movement in the buildup layer, the movement failed to reach the rostral pole. Anderson and coworkers also examined clipped saccade metrics via stimulation of the rostral pole versus saccades clipped via OPN stimulation. These results indicate that there were some differences in the outcomes of the two experiments, most notably, revealing that fixation neurons are not the primary controller for the OPN. These seemingly contradictory pieces of information are easily explained if the fixation pole of the SC is considered an inhibitory unit for the SC only and not the rest of the saccadic system.

While information is shared between the OPN and the fixation pole, this connection operates as a supporting connection only. If the burst from the SC is silenced by the fixation neurons, the signal to the LLBN is clipped and the OPN are released from inhibition. The OPN inhibit the EBN which ends the saccade. Signals sent from the fixation cells to the OPN may serve to reduce the latency between the end of LLBN activity and the resumption of tonic activity. Both Anderson and coworkers and Munoz and coworkers performed experiments to determine the spatial profile of population activity in the buildup layer. Results indicate a buildup layer with very irregular activity profiles that vary from one measurement to the next. A Gaussian function is used to simulate the burst layer to modeling the activity within the deeper layers of the SC prior to, during, and after a saccade (Short and Enderle, 2001).

Within the model by Short and Enderle (2001), feedback is introduced to force the moving hill across the motor map. This feedback comes from the NPH projection to the SC. As the buildup layer receives the current gaze error, the Gaussian peak of the buildup population shifts to the updated gaze error. The buildup layer uses lateral inhibition to control fixation and movement cell activity. As movement cell activity increases, the fixation neurons are inhibited. The location on the collicular surface that correlates with the center of population activity defines the saccade amplitude and direction. Then the cell that receives this input activates neighboring cells. Current motor error is sent back to the buildup layer, moving the hill of activity toward the rostral pole.

The operation of the moving hill in the SC provides a rudimentary control system consisting of a simple feedback circuit. If the FN is lesioned, the SC does terminate the saccade, but the saccade overshoots its destination. The SC termination is due to the influence of the Substantia Nigra and the fixation pole of the buildup layer. With cerebellar control in place, the hill of activity within the buildup layer does not always reach the rostral pole. Indeed, Anderson and coworkers determined there was no strong evidence to support the moving hill hypothesis. Statistical analysis did not show activation of these cells serially ordered from caudal as opposed to in random order. Activity of individual neurons does not correlate well with the resulting saccadic eye movement as individual burst neurons do not always reflect their positions within the retinotopic motor map. However, the response of the entire active population does correlate well with the desired motor error. Therefore, it is not surprising that individual buildup neurons would demonstrate optimal movement fields that seem contrary to the expected activity defined by their location.

Anderson and coworkers noted that oculomotor neurons cease firing approximately 6 ms before the saccade ends. This means that collicular events need to lead eye movements by approximately 8 ms (Anderson et al., 1998). Since the experimental data collected failed to demonstrate this timing sequence, the robustness of the moving hill is challenged unless the operation of the total saccade network is considered. The original moving hill hypothesis proposed that the rostral spread of activity served as a terminator of the saccade. While this mechanism functions in this capacity, the SC provides a less accurate control system, which is superseded by the precise control of the cerebellum. The rostral spread of activity in the SC can serve to terminate a saccade, with less accuracy, if the FN does not terminate the saccade first.

The experiments of Anderson et al. (1998) were restricted to saccades of size 20° or less, while Munoz and Wurtz (1995b) found the strongest evidence for a rostrally directed spread of activity for larger saccades greater than 30°. Such results seem to conflict with the notion of a common control mechanism for saccade termination. One could argue the reason for a caudal to rostral spread of activity is more easily observed for larger saccades is that the active population of buildup neurons is large for any size saccade, and it encompasses a substantial portion of the total collicular motor map. Therefore, individually sampled neurons near the rostral pole would be more likely to be silent at the beginning of buildup activity (100 ms before saccade initiation) or at least, more likely to undergo a more substantial change in activity as the saccade progressed when the initial locus of activity is the farthest from the rostral pole. However, it is true that the more caudal locations have been associated with combined eye-head gaze shifts (Anderson et al., 1998). Port et al. (2000) demonstrated that the rostral spread of activity is not limited to the area corresponding with saccades larger than 20°. Thus, there still remains evidence for the rostral movement of activity.

The SC model described here proposes an explanation of the inconsistencies between the existence of physiologically defined connections between the SC and the PPRF, and the apparent lack of their use (Short and Enderle, 2001). The normal operation of the saccade generator has the cerebellum terminating the saccade before the more inaccurate SC control system ends the inhibition of the OPN.

2.4.2.1 SC Moving Hill Model Description

The model shown in Fig. 2.16 is generated using SIMULINK to simulate saccades and is based on known anatomical connections and physiological interactions. The SC is modeled as three separate subsystems: the superficial, intermediate, and deep layers. The superficial layer acts as a relay of visual information from the retina to the LGN and NRTP. The intermediate layer models the burst layer of the SC and as such receives both excitatory and inhibitory afferent signals and projects to the PPRF. The deep layer receives signals relating desired eye movement, and it converts the polar position coordinates to rectangular collicular coordinates according to the following function:

$$u = Bu(\log(R^2 + A^2 + \sqrt{(2AR\cos(x))})) - Bu(\log(A)) \tag{2.3}$$

$$v = Bv\left(\arctan\left(\frac{R\sin(x)}{(R\cos(x) + A)}\right)\right) \tag{2.4}$$

with
> $Bu = 1.4$
> $Bv = 1.8$
> $A = 3.0$

where u is the horizontal position and v is the vertical position in millimeters. Bu is the scaling constant that determines the size of the collicular map along the u axis (mm), and Bv is the scaling constant that determines the size along the v axis (mm/deg). A is a constant that helps determine the shape of the mapping (deg), and R is the amplitude of the saccade (deg) and x is the direction (deg). Figure 2.17 illustrates the organization of the coordinates with respect to the retinotopic mapping within the SC and one half of the visual field.

The rostral pole is mapped at $u = 0$ mm with the most caudal locations at approximately $u = 4$ mm, based on Ottes et al. (1986) as the general form of the active field. Ottes and coworkers showed the mapping works best when an anisotropic weighing is used, that is, the values of Bu and Bv are not equal.

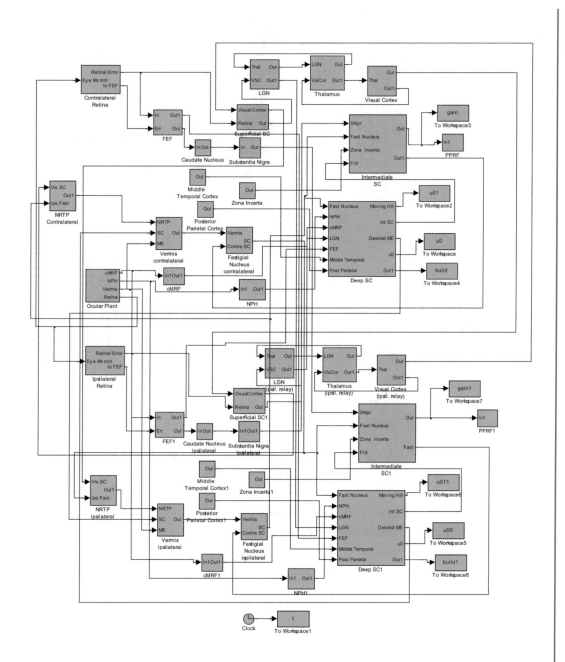

Figure 2.16: Schematic of the Simulink Model for the Superior Colliculus.

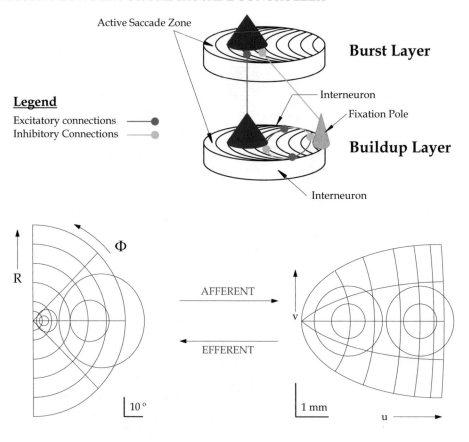

Figure 2.17: (Top) SC Sequence of Activity. (Bottom) Retinotopic Mapping. Based on Ottes et al. (1986).

Ottes and coworkers also developed a general form for the shape of the movement field in the motor population of cells that has the form:

$$W = Wa + Wb\, e^{\frac{-((u-u0)^2+(v-v0)^2)}{2z^2}}$$

(2.5)

with

$Wa = 900$

$Wb = 100$

$z = 0.7$

The values of $u0$ and $v0$ are defined as the components of the vector location of the cell that is central to the active region in the motor neurons for a saccade. These coordinates are found by solving Eqs. (2.3) and (2.4) for a specified saccade amplitude and direction. Wa and Wb are constants that relate the peak firing rate in the population (spikes/s) and the background firing rate (spikes/s), respectively. Wa is defined as $900(u/u0)$ to include the observed phenomenon that the level of activity in the burst layer decreases as the saccade progresses. The term z is a constant that determines the rate of decay of neural activity in cells, with distance from the center $(u0, v0)$, in the superior collicular layer.

The location on the collicular map is related to the burst layer to define the future active site within that layer. Activity within the deep layer begins 100 ms prior to saccade initiation at the site of desired motor error. A fourth-order linear homeomorphic muscle model described in Section 3.6 in Book 1 is used to provide accurate feedback of current eye position as the saccade proceeds. Once the eye movement begins, the deep layer uses a feedback system to model the moving hill (Fig. 2.17, top). As the moving hill approaches the rostral pole, activity within the population of fixation neurons increases. This increases activation of inhibitory interneurons, which then begins to inhibit the activity of the more caudal burst neurons. The intermediate burst layer creates an active region centered at the point of optimal saccade amplitude.

Experiments have shown that the decay rate of activation is not symmetrical with respect to the center of the active region. The level of firing of cells that have optimal amplitudes (considering horizontal movements only) of values less than the current saccade fire more vigorously at the same distance from center as a cell which has an optimal amplitude greater than the intended saccade. However, with adjustment of z, this equation very closely matches the experimental microstimulation data and performs well enough for the burst layer. More recent experiments have found newer methods of quantifying the active saccade zone, but they have still not been able to generate conclusive numbers to accurately quantify the size of the active saccade zone.

After examining the data, we decided to use the value of $z = 0.35$ for saccades of amplitudes $10°$ or higher and $z = 0.18$ for saccades less than $10°$. This is somewhat arbitrary, but it seems the best compromise between the constraints of the mapping equations' ability to define the collicular surface map, lack of a clear trend in the data, and the availability of only a few sites with which to determine the optimal constant value. Additionally, the level of neuronal firing within the active region of the burst layer has been observed to drop as the saccade nears its end. This is a general trend that has not been quantified because sometimes the burst cells cease firing before the end of the saccade. More often they fire even after the eyes have reached the target.

Therefore, we do not have a proven formula for this phenomenon and instead simply induce competition between the inhibition of the fixation cells and the excitation from the buildup layer. As the moving hill approaches the rostral pole, activity within the population of fixation neurons increases. This increases activation of inhibitory interneurons and then begins to end the activity of the more caudal burst neurons. The intermediate burst layer creates an active region centered at the point of optimal saccade amplitude and does not perform computational analysis. This does not mean that we believe that the burst layer performs no role in target selection because this has not been shown conclusively. It is our belief that the greater importance of the burst layer is to provide the LLBN with the large burst that initiates the sequence of activity in the final common pathway. This model leaves the task of target selection to the deeper buildup layer.

While the Gaussian curve defined previously works well for the burst layer, the buildup layer does not have a fairly symmetrical shape. Instead, the observed activity of buildup cells describes a function where the decay of activity for cells with movement fields caudal to the maximally firing cell is very similar to the burst layer. The cells that lie rostral to this central location, however, are not so easily defined. Activity occurs in up to three-quarters of the SC map in the buildup layer, the actual amount activated dependent on size of saccade. Several studies have been done to examine the population activity both spatially and temporally (Munoz and Wurtz, 1995a,b; Port et al., 2000; Anderson et al., 1998). These studies agree that the activity in the buildup layer encompasses a larger portion of the whole map for any size saccade than activity for the same saccade in the burst layer. They also agree that the size of the active zone changes with the amplitude of the saccade. However, Munoz and Wurtz have demonstrated a movement of the locus of activity within the buildup layer as a saccade occurs, so that the active region in the buildup layer moves rostrally across the retinotopic map and excites the fixation neurons, ending the saccade. Anderson et al. (1998) could not find enough evidence to support the hypothesis. They did observe some movement but found that the movement failed to reach the rostral pole. They also tested the metrics of saccades clipped via stimulation of the rostral pole versus saccades clipped via OPN stimulation. The results indicated that there were some differences between them. Therefore, the fixation neurons are not the primary controllers for the OPN. These seemingly contradictory pieces of information are explained if the fixation pole of the SC is considered an inhibitory unit for the SC only and not the rest of the saccadic system. While information is shared between the OPN and the fixation pole, this connection is only a supporting connection. If the burst from the SC is silenced by the fixation neurons, then the signal to the LLBN is clipped and the OPN are released from inhibition. Then the OPN can stop the firing of the EBN and stop the activity of the motoneurons, ending the saccade. Signals sent from the fixation cells to the OPN may serve to reduce the latency between the end of LLBN activity and the resumption of tonic activity, though this is speculation since there is no proof of this.

Both Anderson and Munoz performed experiments to determine the spatial profile of population activity in the buildup layer. Results indicate that the buildup layer displays very irregular activity profiles that vary from one measurement to the next. Therefore, for ease of demonstration, we chose to use the following Gaussian function to describe the activity within the deeper layers of the SC prior to, during, and after a saccade:

$$Ae^{\frac{(-((u-uo)^2+(v-v0)^2))}{(2(z)^2)}} + Be^{\frac{(-((u-u0)^2+(v-v0)^2))}{(2(0.1)^2)}}. \tag{2.6}$$

The value of A in Eq. (2.6) refers to the rise in activity of the buildup cells in the 100 ms prior to saccade initiation and B refers to the decrease in fixation pole activity that results from the inhibitory signal from the movement cells in the buildup layer. The value of $u0$ and $v0$ correspond to the collicular coordinates found with Eqs. (2.3) and (2.4), and refer to the current location of the maximal activity. The value of z is 0.5 for saccades greater than $10°$, but it becomes a function

$z = 0.25(\sqrt{(u^2 + v^2)})$ for saccades less than $10°$. This equation was calculated from available data describing the size of the fixation pole. The value of u along the horizontal meridian that corresponds to the border of the fixation pole is 0.72 mm. Therefore, the above equation is derived for z if we assume a semicircular arc defining the border of the fixation pole of radius 0.72 mm and using the value $z = 0.18$ for the minimum size of the active zone.

Feedback must be introduced to drive the moving hill across the motor map. The feedback comes from the NPH and cMRF (for vertical saccades) projections to the SC. As the buildup layer receives the current gaze error, the Gaussian peak of the buildup population shifts to the new remaining gaze error. The full functioning of the intermediate and deep layers of the SC includes target selection. A lateral inhibition exists within the SC, as described previously. The model implements this process by defining the excitatory and inhibitory profiles for a given location on the retinotopic map. The excitatory profile is a Gaussian curve that describes the level of cellular excitation with respect to distance from the location of excitatory input to the collicular layer. This function is defined as:

$$900 + 100e^{\frac{-((u-u0)^2)+(v-0)^2)}{2(0.35)^2}}. \tag{2.7}$$

This function was derived from the research done by Cannon and Robinson (1987, 1985) and Cannon et al. (1983) in such a manner that when combined with the appropriate inhibitory function:

$$100\left(1 - e^{\frac{-((u-u0)^2+(v-0)^2)}{(2)(0.35)^2}}\right) \tag{2.8}$$

the result is a function that describes the net active profile in the burst layer. These equations define the spatial appearance of the net active population in the burst layer through the inhibition of cells at distant locations to one another. Additionally, the buildup layer uses lateral inhibition to control the fixation and movement cell activity, as described above. As movement cell activity increases, the fixation neurons are inhibited with a latency of 1.6 ms (Munoz and Istvan, 1998). The location on the collicular surface that correlates with the center of population activity defines the saccade amplitude and direction desired. This location is related to the burst layer as the center of the population to activate. The cell that receives the input activates or inhibits neighboring cells according to the weighting functions described above and the net excitation population is generated. Current motor error is fed back to the buildup layer, moving the hill of activity toward the rostral pole. Considering only the operation of the SC, the fixation neurons become active again when the u coordinate of the peak of activity is less than 0.72 mm (Munoz and Wurtz, 1995a,b). Normally, activation of the fixation neurons inhibits the firing of both the burst and buildup neurons. However, it has been observed that the maximum firing rate of the active population of burst cells decreases as the moving hill passes across the SC. Therefore, as the value of the retinal error decreases, the u component of the current error is used to reduce the amplitude of the maximum firing rate, seen in Eq. (2.5). The output of the SC is a time dependent burst function defined as:

$$Q = \frac{(2pif)^2}{(s^2 + (4dpif)s + (2pif)^2)} \tag{2.9}$$

with

$$d = 0.85$$
$$f = 10$$

This signal is sent to the LLBN and results in the desired motor activity along the final common pathway.

2.4.2.2 SC Moving Hill Simulation Results

The model is able to describe the neural activity in any of the modeled nuclei, and is intended to simulate the afferent and efferent connections of the SC with respect to the generation of horizontal saccadic eye movements, as well as demonstrate a possible description of the intercollicular interactions. Thus, it is most important to demonstrate the performance of the SC within the model as the saccade is generated. In the following figures, a series of graphs is presented that depict the change in activation within the buildup layer of the SC as the saccade size is chosen and the saccade proceeds. The u and v axes represent the layer of neural tissue comprising the buildup layer while the z axis represents level of activity in Hz. The graphs shown are only a small sampling of the continuous data available and are intended to demonstrate the trend produced by the model. The time is listed below each graph to define the chronology of the process.

The graphs in Fig. 2.18 clearly demonstrate the initial activation of the fixation pole, the simultaneous rise of the movement neuron activity and fall of the fixation neuron activity, and the resumption of tonic fixation pole firing after saccade termination. Also, note the change in active population as the remaining error decreases below $10°$.

The next set of graphs in Fig. 2.19 follow the same pattern as those on the previous pages, but they, instead, reflect the chronology of activity within the burst layer of the SC. Note that the burst layer does not become active until just before the saccade is initiated because of the removal of the inhibition due to the fixation pole, the FN and substantia nigra, which continue to inhibit the burst layer. The values of the axes are the same as those in Fig. 2.18.

The model produces these results for any saccade size. The graphs in Figs. 2.18 and 2.19 were generated for a $20°$ saccade. Other saccade sizes will have differences in the metrics of the graphs but not the form.

This model was developed to explore the possible purpose of the interactions between cells within the SC. It has long been held that the SC is one of the main neural centers of the saccadic eye movement system. As part of the mesencephalon, the SC is a relatively old portion of the human brain. Therefore, it is entirely possible that the complex interactions between the entire ensemble of nuclei in the eye movement system did not evolve as one single system but as several redundant systems. Any of several anatomical locations in the brain can elicit saccades when stimulated, most notably the FEF and SC. The development of the moving hill hypothesis seems entirely possible as a rudimentary control system consisting of a simple feedback circuit. If the connection between the FN and the SC is destroyed, something not easily done *in vivo*, the SC does in fact start and stop the saccade due to the influence of the substantia nigra and the fixation pole of the buildup layer.

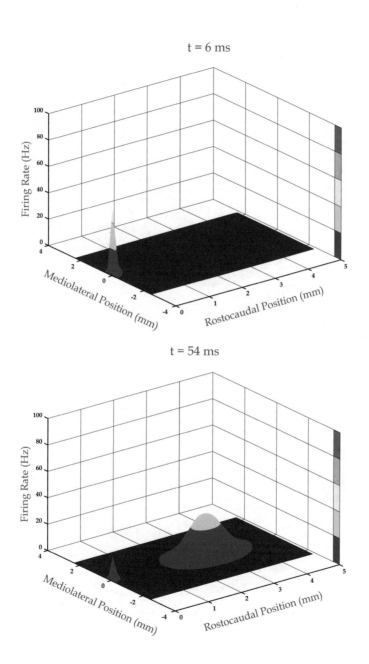

Figure 2.18: a, b. Sequence of population activity within the intermediate layer of the SC.

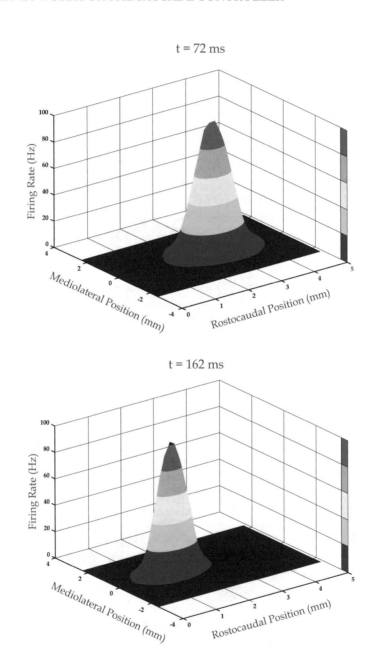

Figure 2.18: c, d. Sequence of population activity within the intermediate layer of the SC.

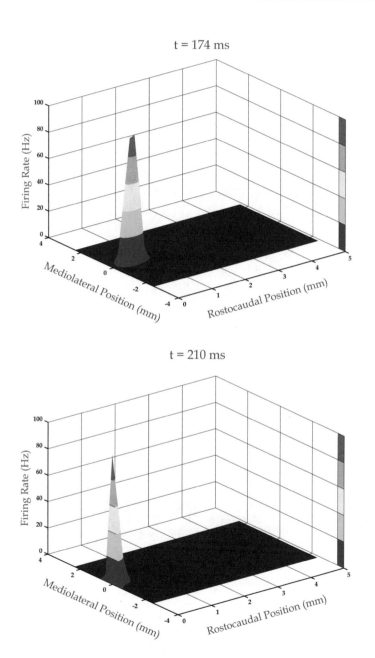

Figure 2.18: e, f. Sequence of population activity within the intermediate layer of the SC.

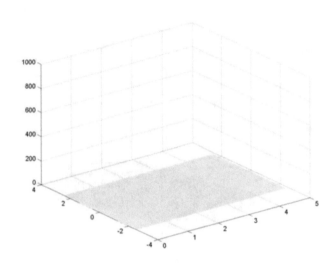

t=0.006 s

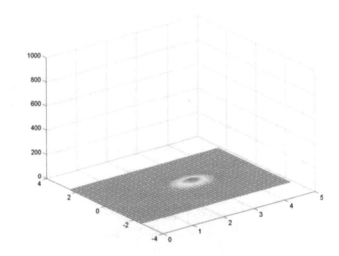

t=0.138 s

Figure 2.19: a, b. Sequence of population activity within the burst layer of the SC.

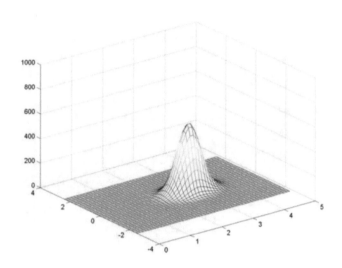

t=0.162 s

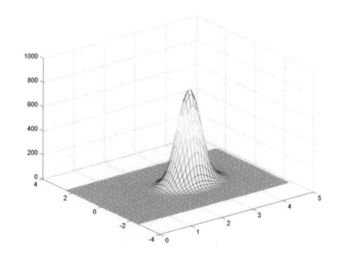

t=0.174 s

Figure 2.19: c, d. Sequence of population activity within the burst layer of the SC.

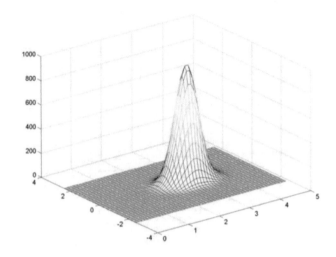

t=0.198 s

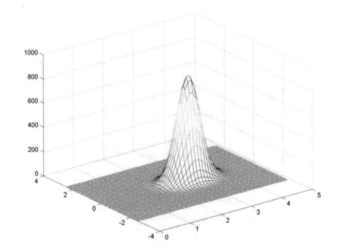

t=0.21 s

Figure 2.19: e, f. Sequence of population activity within the burst layer of the SC.

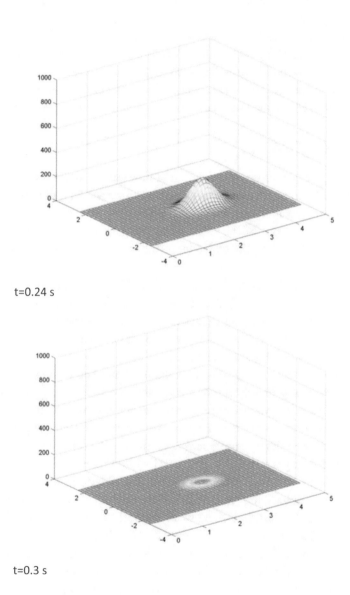

t=0.24 s

t=0.3 s

Figure 2.19: g, h. Sequence of population activity within the burst layer of the SC.

With cerebellar control in place, the hill of activity within the buildup layer does not always reach the rostral pole, which offers an explanation of why there is inconsistency in the observed activity in the rostral pole during a saccade.

This model is a first step toward a much more robust, complete description of a full functioning model of the SC. The lateral inhibition system should be created at the cellular level to better simulate the summation of inputs to the SC including auditory, visual, and remembered targets. The model needs to be expanded to encompass 3-D saccades. The control exerted by the zona incerta is poorly understood and needs to be clarified so that it can be better defined in the model.

2.5   CEREBELLUM

The cerebellum is responsible for the coordination of movements, and it is composed of a cortex of gray matter, internal white matter, and three pairs of deep nuclei: FN, the interposed and globose nucleus, and dentate nucleus as shown in Fig. 2.20. The deep cerebellar nuclei and the vestibular nuclei transmit the entire output of the cerebellum. Output from the cerebellar cortex is carried through Purkinje cells. Purkinje cells send their axons to the deep cerebellar nuclei and have an inhibitory effect on these nuclei.

The cerebellum is involved with both eye and head movements, and both tonic and phasic activity are reported in the cerebellum. The cerebellum is not directly responsible for the initiation or execution of a saccade, but it contributes to saccade precision. Consistent with the operation of the cerebellum for other movement activities, the cerebellum is postulated here to act as the coordinator for a saccade, and act as a precise gating mechanism.

The cerebellum constitutes only 10% of the brain volume, yet it contains more than half of all its neurons. The cerebellum primarily acts as a comparator, compensating for errors by comparing intention with performance via internal and external feedback as shown in Fig. 2.21. Thus, it performs a critical role in coordinating sensory motor information, determining the timing sequence, and the pattern of muscles activated during movement (Hashimoto and Ohtsuka, 1995). The cerebellum is composed of a cortex of gray matter and internal white matter with several areas of the deep cerebellar nuclei (FN, the globise-emboliform nucleus and the dentate nucleus). It receives input from three sources: the periphery, the brain stem, and the cerebral cortex. The input connects to both the cerebellar cortex and the deep nuclei. Most of the outflow from the cerebellar cortex projects back to the deep nuclei (rather than out of the cerebellum) as shown in Fig. 2.21. As a result, neurons in the deep nuclei compare different inputs reaching them directly with the same information after it has been processed by cerebellar cortex. Ito has described this type of connection as providing a precise gaiting mechanism (Ito, M., 1984). The deep cerebellar nuclei are the principal output centers of the cerebellum as a whole.

The cerebellum is included in the saccade generator as a time-optimal gating element, using three active sites during a saccade: the vermis, FN, and flocculus. The vermis is concerned with the absolute starting position of a saccade in the movement field, and it corrects control signals for initial eye position. Using proprioceptors in the oculomotor muscles and an internal eye position reference,

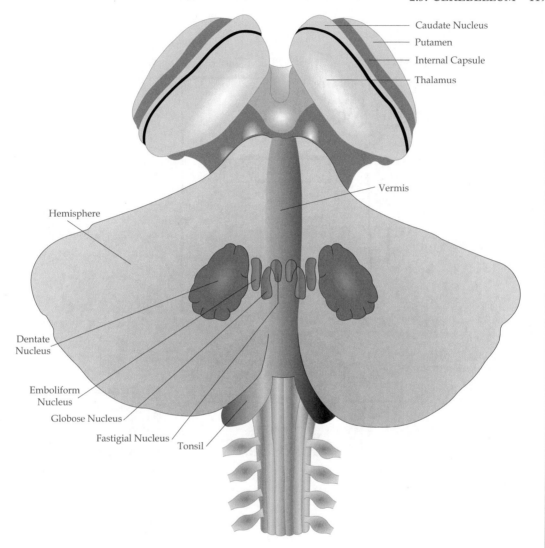

Figure 2.20: An illustration of the cerebellum.

the vermis is aware of the current position of the eye. The vermis is also aware of signals called the dynamic motor error (DME), used to generate the saccade via the connection with the NRTP and the SC.

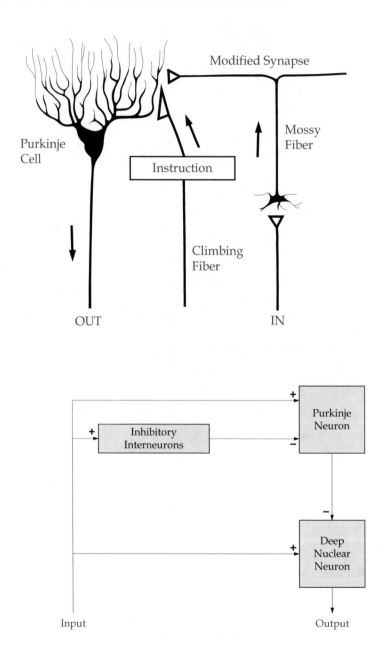

Figure 2.21: (Top) Diagram of important neurons in the cerebellum. (Bottom) Block diagram showing the input-output relationship between the Purkinje and the Deep Nuclear Neuron.

2.5.1   CEREBELLAR STRUCTURE

Each cerebellar fold (folium) is comprised of an outer cerebellar cortex containing three cell layers: molecular, Purkinje, and granule. The cerebellar cortex is a simple and uniform structure consisting of five types of neurons: stellate, basket, Purkinje, Golgi and granule cells as shown in Fig. 2.22. The Purkinje neurons send their axons down through the third layer of the cortex into the underlying whiter matter, and provide the sole output of the cerebellar cortex. In addition to the outflow to the deep nuclei, some portions of the cerebellar cortex project directly to the vestibular nuclei (VN). The cerebellar cortex overlies a deep zone of efferent and afferent fibers that consists of the white matter. Four different zones in the cerebellum receive the afferents: the medial zone, intermediate zone, lateral zone, and flocculonodular zone (or lobe) as shown in Fig. 2.23.

Peduncles, fibers connecting the cerebellum with the brain stem, carry signals in and out of the cerebellum. Inferior cerebellar peduncle contains the dorsal spinocerebellar tract (DSCT) fibers, ventral spinocerebellar tract (VSCT) fibers and corticopontocerebellar pathway. These fibers arise from cells in the ipsilateral Clarke's column in the spinal cord and terminate as mossy fibers. The largest component of the inferior cerebellar peduncle consists of the olivocerebellar tract (OCT) fibers. These fibers arise from the contralateral inferior olive and its fibers terminate in the cerebellum as climbing fibers. Vestibulocerebellar tract (VCT) fibers arise from cells in both the vestibular ganglion and the vestibular nuclei, and they pass in the inferior cerebellar peduncle to reach the cerebellum. Primary vestibular afferents project directly to the ipsilateral cerebellar hemisphere primarily to the flocculonodular lobe. The middle cerebellar peduncle contains the pontocerebellar tract (PCT) fibers. These fibers arise from the contralateral pontine grey neurons that project to the intermediate zone of the cerebellum and receive their corticopontine input from the primary motor cortex. The superior cerebellar peduncle is the primary efferent peduncle of the cerebellum. It contains fibers that arise from deep cerebellar nuclei. These fibers terminate in the red nucleus and the motor nuclei of the thalamus (VA, VL).

The posterolateral fissure on the underside of the cerebellum separates the large posterior lobe from the small flocculonodular lobe. The surface of the cerebellum has two longitudinal furrows that separate three sagittal areas from on another: thin longitudinal strip in the midline (the vermis), and left and right cerebellar hemisphere. The vermis and the hemispheres are connected to different deep cerebellar nuclei. The vermis projects to the fastigial nucleus, and the hemispheres project to the interposed nucleus and the dentate nucleus.

With regard to the oculomotor system, the cerebellum has inputs from SC, LGN, oculomotor muscle proprioceptors, and striate cortex via NRTP. The cerebellum sends inputs to the NRTP, LLBN, EBN, VN, thalamus, and SC as shown in Fig 2.2. The oculomotor vermis and FN are important in the control of saccade amplitude, and the flocculus, perihypoglossal nuclei of the rostral medulla, and possibly the pontine and mesencephalic reticular formation are thought to form the integrator with the cerebellum. One important function of the flocculus may be to increase the time constant of the neural integrator for saccades starting at locations different from primary position.

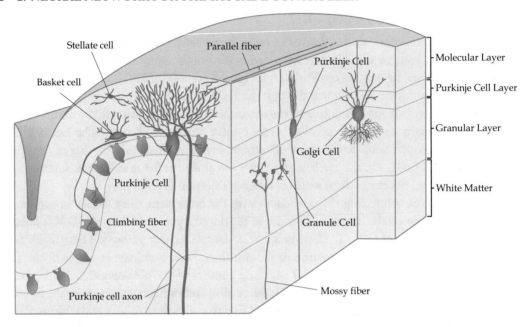

Figure 2.22: A diagram of the five types of neurons: stellate, basket, Purkinje, Golgi and granule cells.

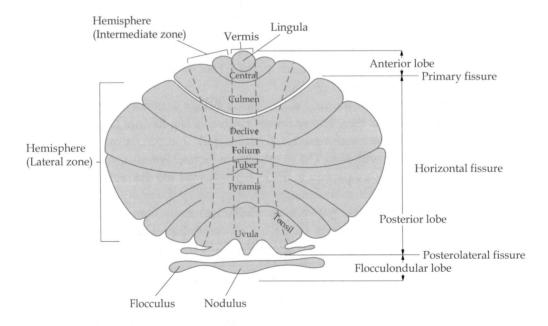

Figure 2.23: Four different zones in the cerebellum receive the afferents: the medial zone, intermediate zone, lateral zone and flocculonodular zone (or lobe).

Purkinje cells exert an inhibitory action on the output of the cerebellum (deep nuclei axons), as shown in Figs. 2.2 and 2.21. As a result, an initial burst of excitation in the deep cerebellar nuclei is followed by inhibition delivered by the Purkinje cells. This mechanism enables the cerebellum to serve as a precise timing device. By means of its output projections to the thalamus and motor cortex and to the red nucleus, the cerebellum regulates the timing necessary to generate the complex patters of muscle activation.

Climbing fibers are axons of neurons from the contralateral inferior olive that go to all parts of the cerebellum. Each climbing fiber sends excitatory collateral to the deep cerebellar nuclei, and then it forms enormous number of synapses with the dendrites of the Purkinje cell, as shown in Fig. 2.22. Each Purkinje cell synapses only with one climbing fiber, but each climbing fiber innervates many Purkinje cells. Climbing fibers give rise to complex spikes in Purkinje cells. Purkinje cells also produce another type of action potential called a simple spike due to innervation of mossy fibers. Unlike climbing fibers, mossy fibers do not go directly to the Purkinje cell. Each mossy fiber forms excitatory synapses on the dendrite of a granule cell. Granule cells have long axons that form parallel fibers. Each parallel fiber forms excitatory synapses with numerous Purkinje cells.

Purkinje cells also have inhibitory input, which come from Golgi cell that are excited by the parallel fibers of granule cells. The axon of the Golgi cell enters into a complex arrangement with the mossy fiber terminal-granule cell dendrite such that the Golgi cell axon inhibits the mossy fiber-granule cell relay (feedback inhibition). Basket cells also provide inhibitory synapses to the somas of the Purkinje cells (feed forward). Purkinje cells distribute their axons in to the four different zones of the cerebellum and project to different deep cerebellar nuclei. Of these nuclei, FN plays an important role in saccadic eye movement system (Hashimoto and Ohtsuka, 1995; Dean, P., 1995; Schweighofer et al., 1996; Versino et al., 1996; Leichnetz and Gonzalo-Ruiz, 1996; Enderle and Engelken, 1996; Goffart and Pelisson, 1997; Krauzlis and Miles, 1998; Takagi et al., 1998). Cells in the FN exit the cerebellum taking two paths. Crossing axons from the FN pass to the vestibular nuclei and reticular formation of the pons and medulla. Uncrossed fibers pass via the inferior cerebellar peduncle to reach the vestibular nuclei and reticular formation.

2.5.2   CEREBELLAR CONTROL OF SACCADES

The cerebellum is involved with eye and head movements. Sites within the cerebellum important for the control of eye movement include the oculomotor vermis, fastigial nucleus, and flocculus. To acquire a target and produce an accurate saccade, two variables are controlled by the CNS. These are the amplitude of the movement (the distance the eye travels) and the direction of the movement. Ito reports that the cerebellum is not directly responsible for the initiation or execution of saccade, but contributes to the saccades precision (Ito, M., 1984).

A portion of the cerebellar cortex, called vestibulocerebellum, is responsible for controlling eye movements and coordinating movements of the head with the eyes. The dominant input to the vestibulocerebellum comes from primary vestibular afferents, originating both in the semicircular canals and the otoliths, and secondary afferents arising from the VN. The vestibulocerebellum also receives visual information from the lateral geniculate nucleus (LGN), SC, and striate cortex. The output is reflected back onto the VN.

The oculomotor vermis is connected with the absolute starting position of a saccade (Krauzlis and Miles, 1998; Takagi et al., 1998; Vilis et al., 1983). The vermis is topographically organized as determined through stimulation studies. It is known that non-primary position saccades have different characteristics than saccades initiated from primary position. For instance, a saccade starting from 30° and moving to primary position has higher peak velocity and shorter duration than a saccade moving from primary position to 30° (this is due to the oculomotor plant elastic elements, etc.). In lesion studies, a saccade is still executed after the cerebellum is removed. However, the accuracy of saccade execution is greatly diminished with marked post saccadic drift (Optican and Miles, 1980). In the model presented here, the saccade in the lesioned case is terminated with the less accurate SC termination circuit, as discussed previously without cerebellar termination. In addition to the cerebellum's role as a coordinator for saccadic eye movements, it is also involved in long term adaptive control (Optican and Miles, 1980, 1985). The deficits in saccade due to damage to the cerebellum result in inaccurate and uncoordinated movements of the eyes. The cerebellum also affects experience dependent modification of saccadic eye movement (Sato and Kawasaki, 1990). Cerebellar control recalibrates the saccade if the controlling muscles undergo a non-catastrophic change over a long period of time.

Saccade-like, rapid eye movements can be evoked by electrical stimulation in restricted areas of the monkey cerebellum, lobules V through VII of the vermis, crus I and II, and the lobulus simples of the hemisphere. The direction of the eye movement is highly dependent on the site of stimulation. When presented on the dorsal aspect of lobules V through VII, the loci for eye movement diverge radially from the intercept of the primary fissure and the mid-sagittal line. Damage to the vermal cortices of lobules VI and VII results in dysmetric saccades. Similarly, dysmetric saccades induced by partial tenotomy of extraocular muscles no longer recover after ablation of lobules VI and VII. In cat lobules VI and VII, Purkinje cells fire 11 to 24 ms before the beginning of saccades, with peak activity falling at the time of onset of eye movement. The intensity of neuronal activity is inversely proportional to the amplitude of the eye movement, suggesting that the Purkinje cells in these areas are involved in adjusting the saccade amplitude. Purkinje cells in lobules VI and VII project to FN and interpositus nuclei. Even though electrical stimulations produced saccade like eye movements, it has not been possible to identify specific neuronal pathways underlying the cerebellar control of saccades.

2.5.3   ROLE OF FASTIGIAL NUCLEUS

The FN receives input from the SC, as well as other sites. The output of the FN is excitatory and projects ipsilaterally and contralaterally, as shown in Fig. 2.2. During fixation, the FN fires tonically at low rates. Twenty ms prior to a saccade, the contralateral FN bursts, and the ipsilateral FN pauses, and then discharges with a burst. The pause in ipsilateral firing is due to Purkinje cell input to the FN. The sequential organization of Purkinje cells along beams of parallel fibers suggests that the cerebellar cortex might function as a delay, producing a set of timed pulses, which could be used to program the duration of the saccade. If one considers nonprimary position saccades, different temporal and spatial schemes, via cerebellar control, are necessary to produce the same size saccade.

Fastigial cells exhibit unique responses depending on the direction of saccades and are involved in terminating the saccade. Purkinje cells in the oculomotor vermis (lobules VIc and VII) are thought to modulate these discharges of fastigial cells (Goffart and Pelisson, 1997; Dean, P., 1995; Sato and Kawasaki, 1990; Sato and Noda, 1992). The cerebellar vermis and its associated deep cerebellar nucleus, the caudal fastigial, is directly implicated in every aspect of the on-line control of saccades: initiation (latency), accuracy (amplitude and direction) and dynamics (velocity and acceleration), and also in the acquisition of adaptive oculomotor behavior. The FN receives topographically organized projections from the anterior and posterior lobe vermis and project bilaterally to the brain stem reticular formation and to the lateral vestibular nuclei. The FN also has crossed ascending projections that reach the motor cortex after relaying in the ventrolateral nucleus.

The exact manner in which the cerebellum intervenes in eye-head coordination is not clear. Those cerebellar areas receiving visual and cervical mossy fiber afferents are connected through the fastigiovestibular projection to the oculomotor system and through the fastigial and/or interpositus nuclei to the tectospinal and tecto-reticulospinal systems, which are responsible for affecting visually triggered head turning. The stability of the gaze can be monitored by vision, which instructs the cerebellar areas via the visual climbing fiber pathway. Thus, it may be inferred that the cerebellum contains corticonuclear microcomplexes that act as sidepaths for the brain stem control for visually triggered saccades and head movement. The relative contribution of these two types of movements could be adjusted by this microcomplex.

2.5.4   CEREBELLAR SACCADE MODEL

Interestingly, experimental data relating the cerebellar vermis and FN to the precise execution of a saccade has resulted in very few mathematical models (Lefevre et al., 1998; Schweighofer et al., 1996). A first order time optimal neural control model for horizontal saccadic eye movement mechanism was proposed by Enderle and Wolfe (1987); Enderle, J. (1994); Enderle et al. (2000). In this model, the control mechanism is initiated by the SC and terminated by the cerebellar FN. Agonist burst cell activity is initiated with maximal firing due to an error between the target and eye position, and it continues until the internal eye position in the cerebellar vermis reaches the desired position, then decays to zero. After the agonist burst, antagonist neural activity rises with a stochastic rebound burst and from input from the FN, then falls to a tonic firing level necessary to keep the eye at its destination. Each of the neural sites in the model fire similarly to experimental data, and each simulate fast eye movements. In 2009, Enderle and coworkers updated the saccade controller (Zhou et al., 2009).

While some of the anatomical connections in the cerebellum are known, many are not regarding to saccades. Future work should attempt to quantitatively model the interaction of Purkinje neurons with the stellate, basket, Golgi, and granule cells, and innervating mossy and parallel fibers, and the FN. The quantitative neuron model should include molecular interactions, and the physiology associated with the dendrite, cell body, axon, and presynaptic terminal. A cerebellar vermis circuit neuron model should be created and be able to quantitatively describe initial eye position regulation.

2.6   SACCADES AND NEURAL ACTIVITY

Neural sites in saccade generator model are described via a *functional* block diagram as shown in Fig. 2.24 (A) and (B). Table 2.2 summarizes additional firing characteristics for the neural sites. The output of each block represents the firing pattern at each neural site observed during the saccade: time *zero* indicates the start of the saccade and *T* represents the end of the saccade. Naturally, the firing pattern observed for each block represents the firing pattern for a single neuron, as recorded in the literature, but the block represents the cumulative effect of all of the neurons within that site. Consistent with a time optimal control theory, neural activity is represented within each of the blocks as pulses and/or steps to reflect their operation as timing gates. Obviously, individual neurons fire, as shown in the figures for the MLBN and LLBN.

In agreement with the first-order time optimal controller, the SC, as described in the neural network figures and the map of its surface, fires as long as the dynamic motor error is greater than zero. Notice that the LLBN are driven by the SC as long as there is a feedback error maintained by the cerebellar vermis. In all likelihood, the firing rate by the SC is stochastic, depending on a variety of physiological factors such as the interest in tracking the target, anxiety, frustration, stress, and other factors.

2.7   TIME OPTIMAL CONTROL OF SACCADES

The saccade generator described here is based on work by Zhou et al. (2009); Enderle, J. (2002); Enderle et al. (2000); Enderle and Engelken (1995); Enderle, J. (1994). The model is first order time optimal, initiated by the intermediate layers of the SC and terminated by the FN of the cerebellum. Under a time optimal saccade controller, the agonist muscle is stimulated by a pulse (the ipsilateral EBN burst) that fires maximally regardless of the size of the saccade, and only the duration of the pulse affects the size of the saccade as shown in Fig. 2.25. In general, whenever a retinal error exists, the contralateral SC fires driving the DME to zero, as described previously with the moving hill, initiating burst activity through the network described in Fig. 2.2 (A). Agonist burst cell activity is initiated with maximal firing due to an error between the target and eye position, and continues until the internal eye position in the cerebellar vermis reaches the desired position, then decays to zero. The cerebellar vermis is responsible for adapting the duration of maximal firing based on the initial position of the eye. After the agonist burst, antagonist neural activity rises with a stochastic rebound burst from FN input, and then it falls to a tonic firing level necessary to keep the eye at its destination. The onset of the antagonist tonic firing is stochastic, weakly coordinated with the end of the agonist burst, and under cerebellar control.

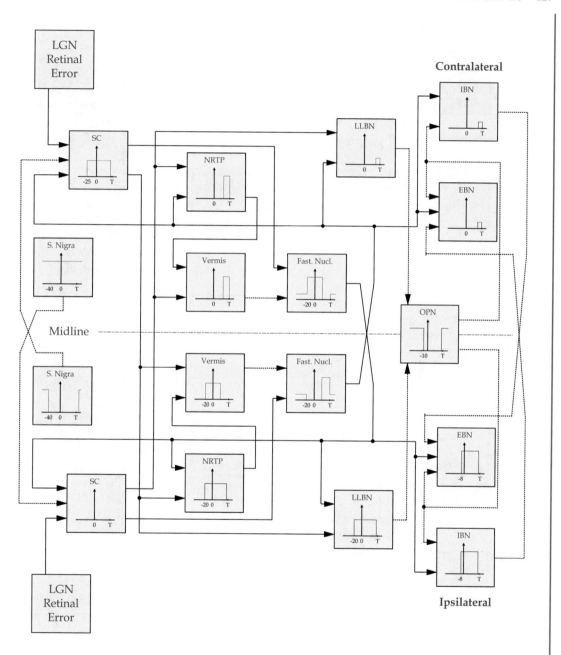

Figure 2.24: (A). A functional block diagram of the saccade generator model. Solid lines are excitatory and dashed lines are inhibitory. This figure illustrates the first half of the network.

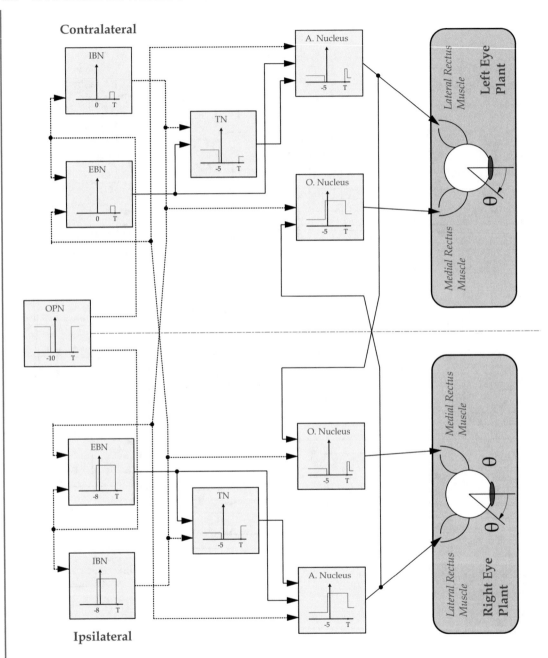

Figure 2.24: (B). A functional block diagram of the saccade generator model. Solid lines are excitatory and dashed lines are inhibitory. This figure illustrates the second half of the network.

Neural Site	Onset Before Saccade	Peak Firing Rate	End Time
Abducens Nucleus	5 ms	400-800 Hz	Ends approx. 5 ms before saccade ends
Contralateral Fastigial Nucleus	20 ms	200 Hz	Pulse ends with pause approx. 10 ms before saccade ends, resumes tonic firing approx. 10 ms after saccade ends
Contralateral Superior Colliculus	20-25 ms	800-1000 Hz	Ends approx. when saccade ends
Ipsilateral Cerebellar Vermis	20-25 ms	600-800 Hz	Ends approx. 25 ms before saccade ends
Ipsilateral EBN	6-8 ms	600-800 Hz	Ends approx. 10 ms before saccade ends
Ipsilateral Fastigial Nucleus	20 ms	Pause during saccade, and a burst of 200 Hz toward the end of the saccade	Pause ends with burst approx. 10 ms before saccade ends, resumes tonic firing approx. 10 ms after saccade ends
Ipsilateral FEF	> 30 ms	600-800 Hz	Ends approx. when saccade ends
Ipsilateral IBN	6-8 ms	600-800 Hz	Ends approx. 10 ms before saccade ends
Ipsilateral LLBN	20 ms	800-1000 Hz	Ends approx. when saccade ends
Ipsilateral NRTP	20-25 ms	800-1000 Hz	Ends approx. when saccade ends
Ipsilateral Substantia Nigra	40 ms	40-100 Hz	Resumes firing approx. 40-150 ms after saccade ends
OPN	6-8 ms	150-200 Hz (before & after)	Ends approx. when saccade ends

Table 2.2: Activity of Neural Sites During a Saccade.

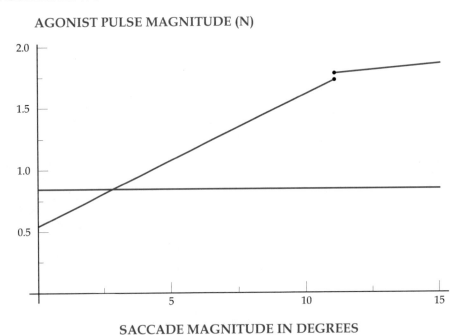

Figure 2.25: Time optimal controller (Blue Line). Amplitude dependent controller based on Bahill, A., 1980 (Red Line).

Both the agonist and antagonist controllers exhibit random behavior from saccade to saccade. This controller is different than others presented in the literature that are either based on a position or velocity model or a vector model. The EBN model presented here supports a time optimal controller in which the firing frequency does not depend on the saccade size and can vary from saccade to saccade of the same size. For saccades above 7°, the saccade size is determined solely by the length of time that the ipsilateral EBN spends bursting in the interval 0 to T2, and the number of neurons firing. As described previously for saccades less than 7°, the duration of the pulse is a constant and the number of neurons firing determines the size of the saccade.

Lefevre et al. (1998) describe two different types of saccade controllers:

(1) Temporal coding of DME by the activity of SC based the dependence of SC burst neuron firing as a function of DME

(2) Spatial coding of DME is reflected via a moving hill in the SC.

Both controllers produce accurate saccades and are based on a local area feedback. The model by Lefèvre and the time optimal model also produce accurate saccades, but they are different than one and two because they use the cerebellum as the terminator of the saccade. A problem with both controllers of type one and two identified by Lefèvre involves lesion studies and the incorrect output of the models. Thus, only the Lefèvre and time optimal models provide a correct output. The Lefèvre model operates by having the SC project to the MLBN directly with excitatory signals dependent on the DME. Further, the Lefèvre model uses the concept of increasing SC projection weights to the MLBN as saccade amplitude increases (a firing rate-saccade amplitude dependent controller), a second order oculomotor plant, and an OPN with a bias similar to the Scudder model. All of the models except for the time optimal model provide an EBN firing rate proportional to the DME. The concept of EBN firing in proportion to the DME is supported indirectly by other studies. However, EBN data that has been reported in the literature does not consistently support this type of controller and even data that is highlighted as supportive does not have a one-to-one relationship between DME and EBN firing rate, nor does it take into account the number of neurons firing.

Chen-Harris et al. (2008) described an adaptive control of saccades via an internal feedback process consistent with the cerebellar control described in Section 2.6. Their model indicated the adaptive control that initiates the saccade changes slowly over time and one that terminates the saccades as changing quickly. The saccade termination is consistent with our theory of cerebellar termination control of saccades.

Generally, saccades recorded for any size amplitude are extremely variable, with wide variations in the latent period, time to peak velocity, peak velocity (indicative of variations in pulse magnitude), and saccade duration. Furthermore, variability is well coordinated for saccades of the same size. Saccades with lower peak velocity are matched with longer saccade durations and saccades with higher peak velocity are matched with shorter saccade durations. Thus, saccades driven to the same destination usually have different trajectories. It is known that saccade dynamics are determined by the number of motoneurons firing in synchrony and their firing rate. Some investigators incorrectly identify the cause of the peak velocity-saccade amplitude profile in the main sequence diagram as due to a nonlinear oculomotor plant driven by a nonlinear saccade amplitude dependent controller. Van Opstal et al. (1985) note "as saccades become larger, their duration increases and peak velocity shows a less-than-linear increase. For a linear system, instead, duration would be the same for all amplitudes and peak velocity would have a linear relation with amplitude."

Shown in the Fig. 2.12 are five saccades elicited from target movements of 4, 8, 12, 16, and 20 degrees. Notice that the 8, 12, 16 and 20 degree eye movements follow the same trajectory before separating toward the end of the eye movement, consistent with a firing rate-independent controller. The 4° eye movement follows a different trajectory and is due to a smaller population of neurons firing maximally, and is time optimal according to the hypothesis presented here. It is envisioned that the reduced number of neurons is based on the "time constant of the burster." That is, once firing is initiated in the EBN, it takes approximately 10 ms to turn off. Thus, small saccades could never occur with all neurons firing maximally, as observed in the microelectrode recordings, if the system did not compensate by reducing the number of active neurons.

Using the system identification technique, Zhou et al. (2009); Enderle and Wolfe (1987) demonstrated that the oculomotor plant is essentially linear, and it does not significantly affect the main sequence diagram. It should be the noted that the main sequence diagram profile is due to the characteristics of the input. Enderle and coworkers (Zhou et al., 2009; Enderle et al., 2000; Enderle et al, 1991) also described a linear muscle model and oculomotor plant that matches the data very closely. Under a time optimal control of saccades, saccade amplitude is determined by the duration of the pulse (interval 0 to T2), whereby the pulse magnitude remains a constant for saccades of all sizes. Given a time optimal input, peak velocity increases in a quasi-linear manner with saccade amplitude up to approximately 15°, after which it reaches a soft saturation level, consistent with the main sequence diagram. For saccades less than 15°, overall maximal peak velocity is not reached because the EBN burst is turned off before this can happen. Note that if the input were a step waveform, then peak velocity would be the same as the point of soft saturation for the pulse-step input at approximately 15°. Because of muscle saturation, these results are unchanged whether the input in Fig. 2.13 (A) or (B) is used.

A firing rate-saccade amplitude dependent controller yields saccades that almost immediately separate into separate trajectories. A firing rate-saccade amplitude independent (time optimal) controller yields saccades, which follow the same trajectory during the pulse phase for saccades over 7°, and then separate. This is usually difficult to appreciate given the large variability in saccades of the same size. A firing rate-saccade amplitude independent (time optimal) controller yields saccades that almost immediately separate into separate trajectories during the pulse phase for saccades under 7° because fewer EBN are firing, as described here. Similar statements can be made for saccade velocity. Note that in Fig. 2.12, the data for position and velocity support a firing rate-saccade amplitude independent (time optimal) controller. In general, families of recorded saccades are not frequently reported in the literature. Of those saccade families reported, Van Opstal et al. (1985) in Fig. 2(a) and Robinson, D. (1964) support saccades with the characteristics like those in Fig. 2.12.

Harwood et al. (1999) attempted to demonstrate that saccades cannot be driven by a rectangular pulse that is time optimal, by examining the spectral main sequence of saccades. Harwood made this conclusion based on the energy spectrum of a rectangular pulse with zeroes at harmonic frequencies of the reciprocal of the pulse duration. Enderle and Wolfe (1987, 1988) demonstrated that the input to the saccade system is a time optimal pulse-step (and not just a pulse) that is low pass filtered, which is not the same as a rectangular pulse bang-bang controller. In Fig. 7 of Enderle and Wolfe (1988), the frequency response of both the data and model prediction is given for a saccade, with the amplitude spectrum showing some minor oscillations in the data. These oscillations in the data are easily explained by sampling and the Gibbs phenomenon. The system identification technique provided model estimates that matched the data extremely well, as illustrated in Fig. 7 of Enderle and Wolfe (1988). Today, based on work with the saccade generator and limitations of the system to switch on and off the pulse as discussed earlier and the new linear homeomorphic oculomotor plant in Section 1.9, we speculate that system identification results will provide an even better match to the data.

Bibliography

Albano, J.E. and Wurtz, R.H., (1981). The role of the primate superior colliculus, pretectum, and posterior-medial thalamus in visually guided eye movements, *Progress in Oculomotor Research*, (edited by Fuchs and Becker). Elsevier, North Holland, pp. 145–153. 96

Anderson R.W., Keller, E.L., Gandhi, N.J., Das, S., (1998). Two dimensional saccade-related population activity in superior colliculus in monkey. *J Neurophysiol* 80: 798–817. 101, 102, 103, 104, 108

Bahill, A.T., Clark, M.R. and Stark, L., (1975). The Main Sequence, A Tool For Studying Human Eye Movements, *Math. Biosci.*, 24: 194–204. DOI: 10.1016/0025-5564(75)90075-9 1

Bahill, A.T. and Hamm, T.M., (1989). Using open-loop experiments to study physiological systems, with examples from the human eye movement systems, *News in Physiol. Sci.*, 4: 104–109.

Bahill, A.T. and Harvey, D.R., (1986). Open-loop experiments for modeling the human eye movement system, *IEEE Trans. Sys. Man. Cybern.*, SMC-16(2): 240–250. DOI: 10.1109/TSMC.1986.4308944

Bahill, A.T. and McDonald, J.D., (1983). Model emulates human smooth pursuit system producing zero-latency target tracking, *Biol. Cyber.*, 48: 213–222. DOI: 10.1007/BF00318089

Bahill, A.T. and McDonald, J.D., (1983b). Frequency limitations and optimal step size for the two-point central difference derivative algorithm with applications to human eye movement data, *IEEE Trans. Biomed. Eng.*, BME-30(3): 191–194. DOI: 10.1109/TBME.1983.325108

Bahill, A.T., *Bioengineering: Biomedical, Medical and Clinical Engineering*, Prentice-Hall, Englewood Cliffs, NJ, 1981.

Bahill, A.T., Brockenbrough A., and Troost, B.T., (1981). Variability and development of a normative data base for saccadic eye movements, *Invest. Ophthal. uis. Sci.*, 21, 116–125.

Bahill, A.T., Latimer, J.R., and Troost, B.T., (1980). Linear homeomorphic model for human movement, *IEEE Trans. Biomed Engr.*, BME-27, 631–639. DOI: 10.1109/TBME.1980.326703 9, 65, 67

Bahill, A.T., (1980). "Development, validation and sensitivity analyses of human eye movement models," *CRC Crit. Rev. Bioeng.*, vol. 4, no. 4, pp. 311–355. 33, 130

Bahill, A.T, Clark, M.R., Stark, L., (1975b). Dynamic overshoot in saccadic eye movements is caused by neurological control signal reversals. *Exp Neurol*, 48: 107–122. DOI: 10.1016/0014-4886(75)90226-5

Becker, W. and Fuchs, A.F., (1985). Prediction in the oculomotor system: smooth pursuit during transient disappearance of visual target. *Experimental Brain Research*, 57: 562–575. DOI: 10.1007/BF00237843

Behan, M. and Kime, N.M., (1996). Intrinsic Circuitry in the Deep Layers of the Cat Superior Colliculus, *Vis. Neurosci.*, 13: 1031–1042. DOI: 10.1017/S0952523800007689 100

Bruce, C.J., Goldberg, M.E., Bushnell, M.C., and Stanton, G.B., (1985). Primate frontal eye fields. II. Physiological and anatomical correlates of electrically evoked eye movements. *Journal of Neurophysiology*, vol. 54, no. 3: 714–34. 98

Bruce, C.J. and Goldberg, M.E., (1985). Primate frontal eye fields, I. single neurons discharging before saccades, *J. Neurophys.*, 53(3): 603–635. 99

Buttner-Ennever, J.A., Horn, A.K., Henn, V., and Cohen, B., (1999). Projections from the superior colliculus motor map to omnipause neurons in monkey, *J. Comp. Neurol.*, Oct 11; 413(1): 55–67. DOI: 10.1002/(SICI)1096-9861(19991011)413:1%3C55::AID-CNE3%3E3.0.CO;2-K 98, 100

Cannon, S.C. and Robinson, D.A., (1987). Loss of the neural integrator of the oculomotor system from brain stem lesions in monkey, *J. Neurophys.*, 57(5): 1383–1409. 109

Cannon, S.C. and Robinson, D.A., (1985). An improved neural-network model for the neural integrator of the oculomotor system: more realistic neuron behavior, *Biol. Cyber.*, 53: 93–108. DOI: 10.1007/BF00337026 109

Cannon, S.C., Robinson, D.A. and Shamma, S., (1983). A proposed neural network for the integrator of the oculomotor system, *Biol. Cyber.*, 49: 127–136. DOI: 10.1007/BF00320393 109

Clark, M.R. and Stark, L., (1975). Time optimal control for human saccadic eye movement, *IEEE Trans. Automat. Contr.*, AC-20: 345–348. DOI: 10.1109/TAC.1975.1100955

Carpenter, R.H.S., *Movements of the Eyes*, 2nd ed., Pion, London, 1988.

Chen-Harris, H., Joiner, W.M., Ethier, V., Zee, D.S., and Shadmehr, R., (2008). Adaptive Control of Saccades via internal Feedback, *J. Neurophys.*, 28(11): 2804–2813. DOI: 10.1523/JNEUROSCI.5300-07.2008 131

Chimoto, S., Iwamoto, Y., Shimazu, H. and Yoshida, K., (1996). Functional connectivity of the superior colliculus with saccade-related brain stem neurons in the cat, *Prog. Brain Res.*, 112: 157–65. DOI: 10.1016/S0079-6123(08)63327-0 82

Close, M.R. and Luff, A.R., (1974). Dynamic properties of inferior rectus muscle of the rat, *J. Physiol.*, London, 236: 258.

Collins, C.C., O'Meara, D. and Scott, A.B., (1975). Muscle tension during unrestrained human eye movements, *J. Physiol.*, 245: 351–369.

Collins, C.C., The human oculomotor control systems, *Basic Mechanisms of Ocular Motility and their Clinical Implications*. G. Lennerstrand and P. Bach-y-Rita, Eds., pp. 145–180. Pergamon Press, Oxford, 1975.

Contreras, D., Destexhe, D., Sejnowski, T. and Steriade, M., (1997). Spatiotemporal Patterns of Spindle Oscillations in Cortex and Thalamus. *J Neurophysiol*, 17(3):1179–1196. 96

Cook, G. and Stark, L., (1968). The human eye movement mechanism: experiments, modeling and model testing. *Archs Ophthal.*, 79, pp 428–436.

Cook, G. and Stark, L., (1967). Derivation of a model for the human eye positioning mechanism, *Bull. Math. Biophys*, 29, 153–174. DOI: 10.1007/BF02476968

Das, S., Gandhi, N.J., and Keller, E.L., (1995). Open-loop simulations of the primate saccadic system using burst cell discharge from the superior colliculus. *Biol. Cybern*, 73:509–518. DOI: 10.1007/BF00199543 101

Dean, P., (1995). Modelling the role of the cerebellar fastigial nuclei in producing accurate saccades: the importance of burst timing, *Neuroscience*, 68(4): 1059–1077. DOI: 10.1016/0306-4522(95)00239-F 123, 125

Descartes, R., *Treatise of Man*. Originally published by Carles Angot, Paris, (1664.) Published translation and commentary by T.S. Hall, Harvard University Press, Cambridge, MA, 1972.

Destexhe, A., Bal, T., McCormick, D.A., Sejnowski, T.J., (1996). Ionic mechanisms underlying synchronized oscillations and propagating waves in a model of ferret thalamic slices. *J Neurophysiol.*, 76:2049 –2070. 96

Enderle, J.D., (1994). A Physiological Neural Network for Saccadic Eye Movement Control. Air Force Material Command, *Armstrong Laboratory AL/AO-TR-1994-0023*: 48 pages. 125, 126

Enderle, J.D., Engelken, E.J., and Stiles, R.N. (1991). A comparison of static and dynamic characteristics between rectus eye muscle and linear muscle model predictions. *IEEE Trans. Biomed. Eng.* 38:1235–1245. 25, 132

Enderle, J.D., *Eye Movements*. In Wiley Encyclopedia of Biomedical Engineering (Metin Akay, Ed.), Hoboken: John Wiley & Sons, 2006.

Enderle, J.D., *The Fast Eye Movement Control System.* In: The Biomedical Engineering Handbook, Biomedical Engineering Fundamentals, 3rd ed., J. Bronzino, Ed., CRC Press, Boca Raton, FL, (2006), Chapter 16, pages 16–1 to 16–21.

Enderle, J.D., Neural Control of Saccades. In J. Hyönä, D. Munoz, W. Heide and R. Radach (Eds.), *The Brain's Eyes: Neurobiological and Clinical Aspects to Oculomotor Research, Progress in Brain Research,* V. 140, Elsevier, Amsterdam, 21–50, 2002. 1, 11, 25, 35, 36, 42, 54, 55, 61, 65, 69, 70, 71, 73, 85, 126

Enderle, J.D., Blanchard, S.M., and Bronzino, J.D., *Introduction to Biomedical Engineering.* Academic Press, San Diego, California, 1062 pages, 2000. 125, 126, 132

Enderle, J.D. and Engelken, E.J., (1996). Effects of Cerebellar Lesions on Saccade Simulations, *Biomed. Sci. Instru.,* 32: 13–22. 123

Enderle, J.D. and Engelken, E.J. (1995). Simulation of Oculomotor Post-Inhibitory Rebound Burst Firing Using a Hodgkin-Huxley Model of a Neuron. *Biomedical Sciences Instrumentation,* 31: 53–58. 1, 84, 85, 126

Enderle, J.D., Engelken, E.J., and Stiles, R.N., (1990). Additional Developments in Oculomotor Plant Modeling. *Biomedical Sciences Instrumentation,* 26: 59–66.

Enderle, J.D. and Wolfe, J.W., (1988). Frequency Response Analysis of Human Saccadic Eye Movements: Estimation of Stochastic Muscle Forces, *Comp. Bio. Med.,* 18: 195–219. DOI: 10.1016/0010-4825(88)90046-7 1, 2, 9, 26, 33, 55, 132

Enderle, J.D., (1988). Observations on Pilot Neurosensory Control Performance During Saccadic Eye Movements, *Aviat., Space, Environ. Med.,* 59: 309–313. 25, 32

Enderle, J.D., and Wolfe, J.W., (1987). Time-Optimal Control of Saccadic Eye Movements. *IEEE Transactions on Biomedical Engineering,* Vol. BME-34, No. 1: 43–55. DOI: 10.1109/TBME.1987.326014 67, 93, 125, 131, 132

Enderle, J.D., Wolfe, J.W., and Yates, J.T. (1984). The Linear Homeomorphic Saccadic Eye Movement Model—A Modification. *IEEE Transactions on Biomedical Engineering,* 1984. Vol. BME-31, No. 11: 711–820. 9

Engelken, E.J., Stevens, K.W., McQueen, W.J. and Enderle, J.D., (1996). Application of Robust Data Processing Methods to the Analysis of Eye Movements, *Biomed. Sci. Instru.,* 32: 7–12.

Engelken, E.J., Stevens, K.W., Bell, A.F. and Enderle, J.D., (1993). Linear Systems Analysis of the Vestibulo-Ocular Reflex: Clinical Applications, *Biomed. Sci. Instru.,* 29: 319–326.

Engelken, E.J., Stevens, K.W. and Enderle, J.D., (1991). Optimization of an Adaptive Non-linear Filter for the Analysis of Nystagmus, *Biomed. Sci. Instru.,* 27: 163–170.

Engelken, E.J., Stevens, K.W. and Enderle, J.D., (1991a). Relationships between Manual Reaction Time and Saccade Latency in Response to Visual and Auditory Stimuli, *Aviat., Space, Environ. Med.,* 62: 315–318. 98

Engelken, E.J., Stevens, K.W. and Enderle, J.D., (1990). Development of a Non-Linear Smoothing Filter for the Processing of Eye-Movement Signals, *Biomed. Sci. Instru.,* 26: 5–10.

Engelken, E.J. and Stevens, K.W., (1989). Saccadic Eye Movements in Response to Visual, Auditory, and Bisensory Stimuli, *Aviat., Space, Environ. Med.,* 762–768. 98

Engelken, E.J., Stevens, K.W., Wolfe, J.W., and Yates, J.T., (1984). A Limbus Sensing Eye-Movement Recorder. Brooks AFB, TX: USAF School of Aerospace Medicine. *USAFSAM-TR-84-29.*

Fenn, W.O. and Marsh, B.S., (1935). Muscular force a different speeds of shortening, *J. Physiol.,* London, 85: 277–297.

Fuchs, A.F., Kaneko, C.R.S., Scudder, C.A. (1985). Brainstem Control of Saccadic Eye Movements. *Ann Rev. Neurosci,* 8:307–337. DOI: 10.1146/annurev.ne.08.030185.001515 55, 65, 82, 85

Fuchs, A.F. and Luschei, E.S., (1971). Development of isometric tension in simian extraocular muscle. *J. Physiol.,* 219, 155–66. 44

Fuchs, A.F. and Luschei, E.S., (1970). Firing patterns of abducens neurons of alert monkeys in relationship to horizontal eye movement. *J. Neurophysiol.,* 33 (3), 382–392. 61

Galiana, H.L., (1991). A Nystagmus strategy to linearize the vestibulo-ocular reflex, *IEEE Trans. Biomed. Eng.,* 38: 532–543. DOI: 10.1109/10.81578 93

Gancarz, G. and Grossberg, S., (1998). A neural model of the saccade generator in reticular formation. *Neural Networks,* vol. 11: pp. 1159–1174. DOI: 10.1016/S0893-6080(98)00096-3 84, 93, 94

Gandhi, N.J. and Keller, E.L., (1997). Spatial distribution and discharge characteristics of the superior colliculus neurons antidromically activated from the omnipause region in monkey. *J. Neurophysiol.,* 76: 2221–5. 76

Girard, B. and Berthoz, A., (2005). From brainstem to cortex: Computational models of saccade generation circuitry. *Progress in Neurobiology,* 77: 215–251. DOI: 10.1016/j.pneurobio.2005.11.001 69

Goffart, L. and Pelisson, D., (1997). Changes in initiation of orienting gaze shifts after muscimol inactivation of the caudal fastigial nucleus in the cat, *J. Physiology,* Sept 15, 503(3): 657–71. DOI: 10.1111/j.1469-7793.1997.657bg.x 123, 125

Goldberg, M.E. and Bushnell, M.C., (1981). Behavioral enhancement of visual responses in monkey cerebral cortex. II. Modulation in frontal eye fields specifically related to saccades. *Journal of Neurophysiology*, vol. 46, no. 4: 773–87. 99

Goldstein, H.P. and Robinson, D.A., (1986). Hysteresis and slow drift in abducens unit activity. *Journal of Neurophysiology*, vol. 55, pp. 1044–1056.

Goldstein, H. and Robinson, D., (1984). A two-element oculomotor plant model resolves problems inherent in a single-element plant model. *Society for Neuroscience Abstracts, 10,* 909.

Goldstein, H., (1983). The neural encoding of saccades in the rhesus monkey (Ph.D. dissertation). Baltimore, MD: The Johns Hopkins University. 2, 9

Graybiel, A.M., (1978). Organization of the nigrotectal connection: an experimental tracer study in the cat. *Brain Research*, vol. 143, no. 2: 339–48. DOI: 10.1016/0006-8993(78)90573-5 99

Harris, C.M. and Wolpert, D.M., (2006). The main sequence of saccades optimizes speed-accuracy trade-off. *Biol Cybern*, 95 (1), 21–29. DOI: 10.1007/s00422-006-0064-x 67

Harting J.K., (1977). Descending pathways from the superior colliculus: an autoradiographic analysis in the rhesus monkey (Macaca mulatta). *J. Comp. Neurol.*, 173: 583–612. DOI: 10.1002/cne.901730311 82

Harwood, M.R., Mezey, L.E., and Harris, C.M., (1999). The Spectral Main Sequence of Human Saccades, *The Journal of Neuroscience*, 19(20): 9098–9106. 132

Hashimoto, M. and Ohtsuka, K., (1995). Transcranial magnetic stimulation over the posterior cerebellum during visually guided saccades in man, *Brain*, 118,(5): 1185–93. DOI: 10.1093/brain/118.5.1185 118, 123

Hikosaka, O. and Wurtz, R.H., The role of substantia nigra in the initiation of saccadic eye movements, *Progress in Oculomotor Research*, (edited by Fuchs and Becker), Elsevier, North Holland, pp. 145–153, 1981.

Hikosaka, O. and Wurtz, R.H., (1983a). Visual and oculomotor functions of monkey substantia nigra pars reticulata. I. Relation of visual and auditory responses to saccades, *J. Neurophys.*, May 49(5): 1230–53. 99

Hikosaka, O. and Wurtz, R.H., (1983b). Visual and oculomotor functions of monkey substantia nigra pars reticulata. II. Visual responses related to fixation of gaze, *J. Neurophys.*, May 49(5): 1254–67. 99

Hill, A.V., (1951a). The transition from rest to full activity in muscles: the velocity of shortening, *Pro. Royal Soc.*, London (B), 138: 329–338. DOI: 10.1098/rspb.1951.0026

Hill, A.V., (1951b). The effect of series compliance on the tension developed in a muscle twitch, *Pro. Royal Soc.*, London (B), 138: 325–329. DOI: 10.1098/rspb.1951.0025

Hill, A.V., (1950a). The development of the active state of muscle during the latent period, *Pro. Royal Soc.*, London (B), 137: 320–329. DOI: 10.1098/rspb.1950.0043

Hill, A.V., (1950b). The series elastic component of muscle, *Pro. Royal Soc.*, London (B), 137: 273–280. DOI: 10.1098/rspb.1950.0035

Hill, A.V., (1938). The heat of shortening and dynamic constants of muscle, *Pro. Royal Soc.*, London (B), 126: 136–195. DOI: 10.1098/rspb.1938.0050

Hodgkin, A.L., Huxley, A.F., and Katz, B., (1952). Measurement of Current-Voltage Relations in the Membrane of the Giant Axon of *Loligo. Journal of Physiology*, 116: 424–448. 85

Horwitz, G.D. and Newsome, W.T., (1999). Separate signals for target selection and movement specification in the superior colliculus, *Science*, May 14; 284(5417): 1158–61. DOI: 10.1126/science.284.5417.1158 100

Hung, G.K. and Ciuffreda, K.J., *Models of the Visual System*. Kluwer Academic/Plenum Publishers, New York, NY, 2002.

Hsu, F.K., Bahill, A.T. and Stark, L., (1976). Parametric sensitivity of a homeomorphic model for saccadic and vergence eye movements, *Comp. Prog. Biomed.*, 6: 108–116. DOI: 10.1016/0010-468X(76)90032-5

Hu, X., Jiang, H., Gu, C., Li, C. and Sparks, D. (2007). Reliability of Oculomotor Command Signals Carried by Individual Neurons. *PNAS*, 8137–8142. DOI: 10.1073/pnas.0702799104 54, 55, 67, 85

Ito, M., (1984). The modifiable neuronal network of the cerebellum, *Jpn. J. of Physiol.*, 34: 781–792. DOI: 10.2170/jjphysiol.34.781 97, 98, 118, 123

Jahnsen, H. and Llinas, R., (1984a). Electrophysiological properties of guinea pig thalamic neurones: an in vitro study, *J. Physiol.*, London, 349: 205–226. 96

Jahnsen, H. and Llinas, R., (1984b). Ionic basis for the electroresponsiveness and oscillatory properties of guinea pig thalamic neurons in vitro, *J. Physiol.*, London, 349: 227–247. 96

Kandel, E.R., Schwartz, J.H., and Jessell, T.M., *Principles of Neural Science: Fourth Edition*. McGraw-Hill, New York, 2000. 99

Kapoula, Z., Robinson, D.A. and Hain, T.C., (1986). Motion of the eye immediately after a saccade. *Exp Brain Res*, 61: 386–394. DOI: 10.1007/BF00239527 1, 2, 85

Keller, E.L., McPeek, R.M. and Salz, T., (2000). Evidence against direct connections to PPRF EBNs from SC in the monkey, *J. Neurophys.*, 84(3): 1303–13. 82, 83

Keller, E.L., The Brainstem. In: *Eye Movements*, edited by Carpenter RHS. London: Macmillan, pp. 200–223, 1991. 82

Keller, E.L., (1989). The cerebellum, *Rev. Oculo. Res.*, 3: 391–411. 98

Kinariwala, B.K., (1961). Analysis of time varying networks, *IRE Int. Convention Rec.*, 4: 268–276.

Krauzlis, R.J., (2005). The control of voluntary eye movements: new perspectives. *The Neuroscientist*, 11 (2),124–137. DOI: 10.1177/1073858404271196 69, 101

Krauzlis, R.J. and Miles, F.A., (1998). Role of the oculomotor vermis in generating pursuit and saccades: effects of microstimulation, *J. Neurophysiolgy*, Oct 80(4): 2046–62. 123, 124

LaCroix, T.P., Enderle, J.D. and Engelken, E.J., (1990). Characteristics of Saccadic Eye Movements Induced by Auditory Stimuli, *Biomed. Sci. Instru.*, 26: 67–78. 98

Lee, C., Roher, W.H. and Sparks, D.L., (1988). Population coding of saccadic eye movements by neurons in the superior colliculus, *Nature*, 332(24): 357–360. DOI: 10.1038/332357a0 96

Lee, E.B., (1960). Mathematical aspects of the synthesis of linear, minimum response-time controllers, *IRE Trans. Automat. Contr.*, AC-5: 283–289. DOI: 10.1109/TAC.1960.1105031

Lefevre, P., Quaia, C., and Optican, L.M., (1998). Distributed model of control of saccades by superior colliculus and cerebellum. *Neural Networks*, vol. 11: pp. 1175–1190. DOI: 10.1016/S0893-6080(98)00071-9 125, 130

Leichnetz, G.R. and Gonzalo-Ruiz, A., (1996). Prearcuate cortex in the Cebus monkey has cortical and subcortical connections like the macaque frontal eye field and projects to fastigial-recipient oculomotor-related brainstem nuclei [published erratum appears in Brain Res. Bull. 1997; 42(1): following III], *Brain Res. Bull.*, 41(1): 1–29. DOI: 10.1016/0361-9230(96)00154-2 123

Leigh, R.J. and Zee, D.S., *The Neurology of Eye Movements*. Oxford University Press, New York, NY, 1999. 98

Lehman, S. and Stark, L., (1982). Three algorithms for interpreting models consisting of ordinary differential equations: sensitivity coefficients, sensitivity functions, global optimization, *Math. Biosci*, 62, 107–122. DOI: 10.1016/0025-5564(82)90064-5

Lehman, S. and Stark, L., (1979). Simulation of linear and nonlinear eye movement models: sensitivity analyses and enumeration studies of time optimal control, *J. Cybernet. Inf Sci*, 2, 21–43.

Ling, L., Fuchs, A., Siebold, C. and Dean, P., (2007). Effects of initial eye position on saccade-related behavior of abducens nucleus neurons in the primate. *J. Neurophysiol.*, 98 (6), 3581–3599. DOI: 10.1152/jn.00992.2007 61, 85

Markham, C.H., (1981). Cat medial pontine neurons in vestibular nystagmus, *Annals New York Academy of Science*, 374: 189–209. DOI: 10.1111/j.1749-6632.1981.tb30870.x 93

May, A., (1985). An Improved Human Oculomotor Model For Horizontal Saccadic (Fast) Eye Movements, MS. Thesis, North Dakota State University, Fargo, ND.

Meredith, M.A. and Ramoa, A.S., (1998). Intrinsic circuitry of the superior colliculus: pharmacophysiological identification of horizontally oriented inhibitory interneurons, *J. Neurophys.*, Mar 79(3): 1597–602. 100

Miura, K. and Optican, L., (2006). Membrane Chanel Properties of Premotor Excitatory Burst Neurons May Underlie Saccade Slowing After lesions of Ominpause Neurons. *J. Comput. Neurosci.*, 20: 25–41. DOI: 10.1007/s10827-006-4258-y 70, 73, 84

Moschovakis, A.K., Scudder, C.A. and Highstein, S.M., (1996). The Microscopic Anatomy and Physiology of the Mammalian Saccadic System, *Prog. Neurobiol.*, Oct 50(2–3): 133-254. DOI: 10.1016/S0301-0082(96)00034-2 82, 83, 93, 98, 99

Munoz, D.P. and Istvan, P.J., (1998). Lateral inhibitory interactions in the intermediate layers of the monkey superior colliculus, *J. Neurophys.*, Mar 79(3): 1193–209. 100, 102, 109

Munoz, D.P., Wurtz, R.H., (1995a). Saccade-related activity in monkey superior colliculus. I. Characteristics of burst and buildup cells. *J. Neurophysiol.*, 73: 2313–2333. 99, 100, 101, 102, 108, 109

Munoz, D.P. and Wurtz, R.H., (1995b). Saccade-related activity in monkey superior colliculus. II. Spread of activity during saccades. *J. Neurophysiol.*, 73: 2334–2348. 99, 100, 101, 102, 103, 108, 109

Munoz, D.P. and Wurtz, R.H., (1993a). Fixation cells in the monkey superior colliculus. I. Characteristics of cell discharge, *J. Neurophys.*, 70(2): 559–575. 99, 102

Munoz, D.P. and Wurtz, R.H., (1993b). Fixation cells in the monkey superior colliculus. II. Reversible activation and deactivation, *J. Neurophys.*, 70(2): 576–589. 99, 102

Munoz, D.P. and Wurtz, R.H., (1992). Role of the Rostral Superior Colliculus in Active Visual Fixation and Execution of Express Saccades, *J. Neurophys.*, 67(4): 1000–1002. 99, 102

Munoz, D.P., Pelisson, D. and Guitton, D., (1991). Movement of Neural Activity on the Superior Colliculus Motor Map During Gaze Shifts, *Science*, 251(4999): 1358–60. DOI: 10.1126/science.2003221 101, 102

Nakahara, H., Morita, K., Wurtz, R.H., and Optican, L.M., (2006). Saccade-Related Spread of Activity Across Superior Colliculus May Arise From Asymmetry of Internal Connections. *J. Neurophysiol.*, 96: 765–774. DOI: 10.1152/jn.01372.2005 96

Ohki, Y., Shimazu, H., and Suzuki, I., (1988). Excitatory input to burst neurons from the labyrinth and its mediating pathway in the cat: location and functional characteristics of burster-driving neurons. *Exp. Brain Res.*, 72: 457–472. DOI: 10.1007/BF00250591 93

Ohtsuka, K. and Nagasaka, Y., (1999). Divergent axon collaterals from the rostral superior colliculus to the pretectal accommodation-related areas and the omnipause neuron area in the cat, *J. Comp. Neurol.*, 11; 413(1): 68–76,. DOI: 10.1002/(SICI)1096-9861(19991011)413:1%3C68::AID-CNE4%3E3.0.CO;2-7 98, 100

Olivier, E., Grantyn, A., Chat, M. and Berthoz, A., (1993). The control of slow orienting eye movements by tectoreticulospinal neurons in the catbehavior, discharge patterns and underlying connections. *Exp. Brain Res.*, 93: 435–449. DOI: 10.1007/BF00229359 82

Optican, L.M. and Miles, F.A., (1985). Visually induced adaptive changes in primate saccadic oculomotor control signals. *J. Neurophysiol.*, 54 (4), 940–958. 2, 124

Optican, L.M. and Miles, F.A., (1980). Cerebellar-dependent adaptive control of primate saccadic system, *J. Neurophys.*, 44(6): 1058–1076. 124

Ottes, F.P., Van Gisbergen, J.A.M. and Eggermont, J.J., (1986). Visuomotor Fields of the Superior Colliculus: A Quantitative Model, *Vis. Res.*, 26(6): 857–873. DOI: 10.1016/0042-6989(86)90144-6 96, 104, 106

Pierre, D.A., *Optimization Theory with Application*, Wiley, New York, pp. 277–280, 1969.

Port, N.L., Sommer, M.A. and Wurtz, R.H., (2000). Multielectrode evidence for spreading activity across the superior colliculus movement map, *J. Neurophys.*, Jul 84(1): 344–57. 102, 104, 108

Quaia, C. and Optican, L.M., (1998). Commutative saccadic generator is sufficient to control a 3-D ocular plant with pulleys. *J. Neurophysiol.*, 79 (6), 3197–3215. 1

Quaia, C. and Optican, L.M., (2003). Dynamic eye plant models and the control of eye movements. *Strabismus*, 11(1), 17–31. DOI: 10.1076/stra.11.1.17.14088 1

Ramat, S., Leigh, R.J., Zee, D. and Optican, L., (2007). What Clinical Disorders Tell Us about the Neural Control of Saccadic Eye Movements. *Brain*, 1–26. DOI: 10.1016/j.ajo.2007.01.007 76

Ramat, S., Leigh, R.J., Zee, D. and Optican, L., (2005). Ocular Oscillations generated by Coupling of Brainstem Excitatory and Inhibitory Saccadic Burst Neurons. *Exp. Brain Res.*, 160: 89–106. DOI: 10.1007/s00221-004-1989-8 70, 73

Raybourn, M.S. and Keller, E.L., (1977). Colliculoreticular organization in primate oculomotor system. *J. Neurophysiol.*, 40: 861–878. 82, 96

Ricardo, J.A., (1981). Efferent connections of the subthalamic region in the rat. II. The zona incerta, *Brain Res.*, Jun 9; 214(1): 43–60. DOI: 10.1016/0006-8993(81)90437-6 99

Robinson, D.A., Models of mechanics of eye movements. In: B.L. Zuber (Ed.), *Models of Oculomotor Behavior and Control* (pp. 21–41). Boca Raton, FL: CRC Press, 1981. 2, 9, 31, 39, 61, 66, 85, 87, 88, 93, 94, 98

Robinson, D.A., (1973). Models of the saccadic eye movement control system. *Kybernetik,* 14: 71–83. DOI: 10.1007/BF00288906

Robinson, D.A., (1972). Eye movements evoked by collicular stimulation in the alert monkey, *Vis. Res.*, 12: 1795–1808. DOI: 10.1016/0042-6989(72)90070-3 96

Robinson, D.A., O'Meara, D.M., Scott, A.B. and Collins, C.C., (1969). Mechanical components of human eye movements, *J. Appl. Physiol.* 26, 548–553.

Robinson, D.A., (1964). The Mechanics of Human Saccadic Eye Movement, *J. Physiol.*, London, 174: 245. 132

Sato, H. and Noda, H., (1992). Saccadic dysmetria induced by transient functional decoration of the cerebellar vermis, *Brain Res. Rev.*, 8(2): 455–458. DOI: 10.1007/BF02259122 125

Sato, Y. and Kawasaki, T., (1990). Operational unit responsible for plane-specific control of eye movement by cerebellar flocculus in cat, *J. Neurophys.*, 64(2): 551–564. 124, 125

Schweighofer, N., Arbib, M.A. and Dominey, P.F., (1996). A model of the cerebellum in adaptive control of saccadic gain. II. Simulation results, *Biol. Cyber.*, Jul 75(1): 29–36. DOI: 10.1007/BF00238737 123, 125

Scudder, C.A., Kaneko, C., Fuchs, A., (2002). The brainstem burst generator for saccadic eye movements: a modern synthesis. *Exp. Brain Res.,* 142: 439–462. DOI: 10.1007/s00221-001-0912-9 69, 84

Seidel, R.C., (1975). Transfer-function-parameter estimation from frequency-response data ~ a FORTRAN program, NASA TM X-3286 Report.

Short, S.J. and Enderle, J.D., (2001). A Model of the Internal Control System Within the Superior Colliculus, *Biomed. Sci. Instru.*, 37: 349–354. 96, 102, 103, 104

Smith, Jr., F.W., (1961). Time-optimal control of higher-order systems, *IRE Trans. Automat. Contr.*, vol. AC-6, pp. 16–21.

Sparks, D.L., (2002). The Brainstem Control of Saccadice Eye Movements. *Neuroscience*, 3: 952–964. 69

Sparks, D.L. and Hartwich-Young, R., (1989). The Deep Layers of the Superior Colliculus. *Rev. Oculomot. Res.*, 3: 213–55. 99

Sparks, D.L. and Nelson, J.S., (1987). Sensory and motor maps in mammalian superior colliculus. *TINS*, vol. 10, no. 8: 312–317. DOI: 10.1016/0166-2236(87)90085-3 96

Sparks, D.L., (1986). Translation of Sensory Signals Into Commands for Control of Saccadic Eye Movements: Role of Primate Superior Colliculus, *Physiol. Rev.*, Jan 66(1): 118–171. 96, 97, 98

Sparks, D.L., (1978). Functional Properties of Neurons in the Monkey Superior Colliculus: Coupling of Neuronal Activity and Saccade Onset, *Brain Res.*, 156: 1–16. DOI: 10.1016/0006-8993(78)90075-6 99

Sparks, D.L., Holland, R. and Guthrie, B.L., (1976). Size and Distribution of Movement Fields in the Monkey Superior Colliculus, *Brain Res.*, 113: 21–34. DOI: 10.1016/0006-8993(76)90003-2 10, 48, 80, 81, 83, 92, 99, 101

Stanton, G.B., Goldberg, M.E. and Bruce C., (1988). Frontal eye field efferents in the macaque monkey. I. Subcortical pathways and topography of striatal and thalamic terminal fields. *J. Comp. Neurol.*, 271: 473–492. DOI: 10.1002/cne.902710402 82

Stechison, M.T., Saint-Cyr, J.A. and Spence, S.J., (1985). Projections from the nuclei prepositus hypoglossi and intercalatus to the superior colliculus in the cat: an anatomical study using WGA-HRP. *Experimental Brain Research*, vol. 59, no. 1: 139–50. DOI: 10.1007/BF00237674 98

Sylvestre, P.A. and Cullen, K.E., (2006). Premotor Correlates of Integrated Feedback Control for Eye–Head Gaze Shifts. *J. Neurophysiol.*, 26(18):4922–4929. DOI: 10.1523/JNEUROSCI.4099-05.2006

Sylvestre, P.A. and Cullen, K.E., (1999). Quantitative analysis of abducens neuron discharge dynamics during saccadic and slow eye movements. *J. Neurophysiol.*, 82 (5), 2612–2632. 2, 61, 83, 85

Takagi, M., Zee, D.S. and Tamargo, R.J., (1998). Effects of lesions of the oculomotor vermis on eye movements in primate: saccades, *J. Neurophysiology*, Oct 80(4): 1911–1931. 123, 124

Van Gisbergen, J.A., Robinson, D.A. and Gielen, S. (1981). A quantitative analysis of generation of saccadic eye movements by burst neurons. *J. Neurophysiol.*, 45 (3), 417–442. 2, 9, 61, 66, 85

Van Opstal, A.J., Van Gisbergen, J.A.M. and Eggermont, J.J., (1985). Reconstruction of Neural Control Signals for Saccades Based on a Inverse Method, *Vis. Res.*, 25(6): 789–801. DOI: 10.1016/0042-6989(85)90187-7 131, 132

Versino, M., Hurko, O. and Zee, D.S., (1996). Disorders of binocular control of eye movements in patients with cerebellar dysfunction, *Brain,* Dec 119 (Pt 6): 1933–1950. DOI: 10.1093/brain/119.6.1933 123

Vilis, T., Snow, R. and Hore, J., (1983). Cerebellar saccadic dysmetria is not equal in the two eyes, *Exp. Brain Res.,* 51: 343–350. DOI: 10.1007/BF00237871 124

Vossius, G., (1960). The system of eye movement. *Z. Biol.,* vol. 112: 27–57.

Weber, R.B. and Daroff, R.B., (1972). Corrective movements following refixation saccades: Type and control system analysis. *Vision Res.,* 12:467–475. 1

Westheimer, G., (1954). Mechanism of saccadic eye movements. *AMA Archives of Ophthalmology,* 52: 710–724. DOI: 10.1016/0042-6989(72)90090-9

Westine, D.M. and Enderle, J.D., (1988). Observations on neural activity at the end of saccadic eye movements. *Biomedical Sciences Instrumentation,* 24: 175–185. 1

Widrick, J.J., Romatowski, J.G., Karhanek, M. and Fitts, R.H., (1997). Contractile properties of rat, rhesus monkey, and human type I muscle fibers. *Am. J. Physiol.,* 272, R34–42. 44, 45

Wolfe, J.W., Engelken, E.J., Olson, J.W. and Allen, J.P., (1978). Cross-power spectral density analysis of pursuit tracking. *Annals of Otology, Rhinology and Laryngology,* 87(6): 837–844.

Wurtz, R.H. and Goldberg, M.E., (1972). Activity of superior colliculus in behaving monkey. III. Cells discharging before eye movements. *Journal of Neurophysiology,* vol. 35: 575–586. 96

Yasui, S. and Young, L.R., (1975.) Perceived visual motion as effective stimulus to pursuit eye movement system. *Science,* 190: 906–908.

Zee, D.S., Optican, L.M., Cook, J.D., Robinson, D.A. and Engel, W.K., (1976). Slow saccades in spinocerebellar degeneration, *Arch. Neurol.,* vol. 33, pp. 243–251. DOI: 10.1126/science.1188373

Zhou, W., Chen, X. and Enderle (2009). An Updated Time-Optimal 3rd-Order Linear Saccadic Eye Plant Model. *International Journal of Neural Systems,* In Press. 1, 2, 11, 25, 69, 70, 71, 84, 125, 126, 131, 132

Authors' Biographies

JOHN D. ENDERLE

John D. Enderle, Biomedical Engineering Program Director and Professor of Electrical & Computer Engineering at the University of Connecticut, received the B.S., M.E., and Ph.D. degrees in biomedical engineering, and M.E. degree in electrical engineering from Rensselaer Polytechnic Institute, Troy, New York, in 1975, 1977, 1980, and 1978, respectively.

Dr. Enderle is a Fellow of the IEEE, the past Editor-in-Chief of the *EMB Magazine* (2002-2008), the 2004 EMBS Service Award Recipient, Past-President of the IEEE-EMBS, and EMBS Conference Chair for the 22nd Annual International Conference of the IEEE EMBS and World Congress on Medical Physics and Biomedical Engineering in 2000. He is also a Fellow of the American Institute for Medical and Biological Engineering (AIMBE), Fellow of the American Society for Engineering Education and a Fellow of the Biomedical Engineering Society. Enderle is a former member of the ABET Engineering Accreditation Commission (2004-2009). In 2007, Enderle received the ASEE National Fred Merryfield Design Award. He is also a Teaching Fellow at the University of Connecticut since 1998.

Enderle is also involved with research to aid persons with disabilities. He is Editor of the NSF Book Series on *NSF Engineering Senior Design Projects to Aid Persons with Disabilities*, published annually since 1989. Enderle is also an author of the book *Introduction to Biomedical Engineering*, published by Elsevier in 2000 (first edition) and 2005 (second edition). Enderle's current research interest involves characterizing the neurosensory control of the human visual and auditory system.

WEI ZHOU

Wei Zhou holds two bachelor's degrees in mechanical engineering (safety engineering) and electrical engineering from the University of Science and Technology of China, and an MS degree in biomedical engineering from the University of Connecticut. He is pursuing the PhD degree in biomedical engineering at the University of Michigan. His research area involves biomechanics, motor control, and ergonomics.